Everyone Has A Story

Everyone Has A Story

A Combat Medic's Tale

RANDALL W. ARK

Deeds Publishing | Athens

Published by Deeds Publishing in Athens, GA
www.deedspublishing.com

Printed in The United States of America

Cover and interior design by Deeds Publishing

ISBN 978-1-961505-32-2

Books are available in quantity for promotional or premium use. For information, email info@deedspublishing.com.

First Edition, 2024

10 9 8 7 6 5 4 3 2 1

dedication to go here

Contents

Preface   ix

One Soldier's Story   1
    A Little Background   1
    Preparing for Basic Training   2
    Close Calls   3
    Flying to Ft. Jackson, April 4, 1968   4
    The Fort Jackson Reception Station   4
    Boot Camp   8
    Ft. Sam Houston, San Antonio, Texas   15
    Heading home for 30 days leave   20
    Leaving Home   20

Finding Jesse   57

Additional Images   129

Acknowledgements   141

About the Author   143

Preface

To Whomever Chooses to Read This Book:

At the request of a few friends and WWII veterans whom I frequently visit, I decided to write of my experiences in Vietnam and the subsequent trials of coming home. Each veteran has his or her own story, and one person's story is neither better nor worse than another's, nor any more important. This is merely my story.

Too many veterans will die and leave behind loved ones who know virtually nothing of what they experienced or saw, or how they felt about what they did. My own family, my close friends, and my children have little knowledge of where I was, what I did, and how I felt about my tour in Vietnam from September 1968 to late August 1969.

As a combat veteran, I am forced to sort out many varieties of feelings in a way that enables me to function in society as a productive citizen. Along with the feelings of gratefulness and an appreciation for life, I live with survival guilt, nightmares, flashbacks, and a host of other "side effects." Writing about my experience has helped, and I pray that it might help someone else.

Blessings,
Randall W. Ark

One Soldier's Story

A LITTLE BACKGROUND

I was raised on a small farm outside of Springfield, Ohio, and I attended Greenon High School, graduating in 1966. I turned 18 just after graduating.

I began college in the fall at Wright State College in Dayton, Ohio. Wright State had not yet reached university status as there were only four buildings on campus at that time. The lottery system had not yet been put into place and I was on academic probation so many times that I finally heard from my Uncle Sam; I was classified 1-A and drafted into the military in November of 1967.

To tell you the truth, I was relieved. I was tired of the rigors of school life, and I wanted to do something different. I wanted to be on my own and live somewhere else. It's not that I had a bad homelife, I had a great homelife and a great family. I just needed a change, and boy, change is what I got.

One thing I discovered is that if you were drafted, you only had to serve in the military for two years and your serial number began with "US.", but you had very little say-so as to where you might be assigned or how you would be trained. But if you *joined* the service (enlisted) it was for three years, and your serial number would begin with "RA," but you could choose what you wanted to do. I thought about it, and I decided I wanted to be a medical corpsman, a medic. My thinking was that even if I went to Vietnam, I would be safely working in an air-conditioned hospital somewhere out of harm's way and surrounded by pretty nurses. What can I say? I was eighteen and ignorant.

To this day, I don't know why someone didn't try to talk me out of that decision. Certainly, my father knew what the chances were for medics in combat. Maybe I wasn't as well-liked as I had led myself to believe. Just kidding. The truth is, my father was not the type to dissuade us from decisions like these, so he left it up to me as to what I wanted to do in the Army.

PREPARING FOR BASIC TRAINING

After being drafted, I was required to have a physical exam at Ft. Hayes, Columbus. I had a cast on my right arm at the time because of a cracked wrist from helping a friend load pigs onto a trailer on a cold Saturday morning. When I passed the physical, I questioned a sergeant there about my cracked wrist, and he said that Charlie Cong would take care of that for me. Such compassion I have rarely seen!

When I went to Ft. Hayes a second time, I was to leave from there to my next duty station which would be Ft. Jackson, South Carolina.

There were three of my Greenon classmates, Jim Hunter, John King, and Charles Swaney, who came and took me out to eat before I caught my plane to boot camp. That was very thoughtful of them to do that. A kindness I'll never forget.

Ft. Jackson, South Carolina

CLOSE CALLS

There were a couple of instances where having joined, instead of being drafted, paid off. At Fort Hayes, Columbus, where everyone had to go for a physical examination, once you had enlisted or had been drafted, I was standing in line after we finished getting our physicals, when a sergeant told us to count off by two's. We did that and then we were told that all the one's were in the Army, and all the two's were told they were in the Marines. Those were the only two choices that a person had in 1968, if he wanted to join the service. All other options were closed, unless you "knew" someone.

Anyway, I counted off as a "two," a Marine, until I told them that I had enlisted in the Army after being drafted, I was an "RA.".

I was then told to get in line with the one's. The big reason I chose joining the Army over the Marines was because of my father. He had served in the Army during WWII, and I wanted to do what he did. Knowing that my dad had gone through what I was about to go through helped me to endure the training and what followed.

FLYING TO FT. JACKSON, APRIL 4, 1968

I had never flown on a plane before, and this plane was a turbo-jet. When it took off, I threw up, and when it landed, I threw up. Unbeknownst to me, classical conditioning was taking place on that plane. Smoking was allowed on planes back then, so the smell of smoke and my sick vomiting paired up, so consequently, I didn't have a desire to smoke all through Basic Training (8 weeks) and all through (AIT) Advanced Individualized training (10 weeks).

Most of the guys on this flight were blacks and Puerto Ricans out of Harlem and when an announcement was broadcast on our plane that Martin Luther King, Jr. had been assassinated, everyone got quiet. And remained quiet. This was April 4, 1968.

THE FORT JACKSON RECEPTION STATION

After landing at the Columbia, South Carolina Airport, we were transported to the Ft. Jackson Reception Station. And what a reception it was. It was at night, and we all were told to file into a building where we would get our new GI haircuts, so we all got buzzed to the scalp. It's like we were standing in line to become a different person, and I guess we were, in reality.

The image on the right was taken in Columbus, Ohio
the night before flying to Ft. Jakcson, SC.

You know, you don't realize it so much until it happens, that when everyone's hair is cut the same, it takes away a whole lot of pre-conceived notions you might have had about another person and the stereotyping of each other. We humans often identify and judge each other by how we wear our hair or the kind of clothes we wear. The Army wants us to look the same, dress the same, respond to orders the same, and become a unified machine, to react as one, to commands. And we were given all this clothing, every article of clothing you can imagine. OD green, of course. Our boots were not jungle boots, but black leather. Our boots and dress shoes were to be spit-shined whenever required, which was often. I got pretty good at spit shining.

The next time my being enlisted paid off was when we were assembled outside of the Reception Station.

We were all called outside after getting all our free stuff and were told to line up at attention. Names were then called out, and my name was called. We who were called were told that we had been selected to train at a helicopter machine gunner school, and that we were going to skip basic training. WHAT?? SKIP BASIC TRAINING? Now, this really caught me off guard. At first it sounded appealing to skip basic training, then logic took over. The helicopter door gunners in Vietnam must be dropping like flies if they are taking new troops right from the Reception Station to the machine gunner school.

I immediately told the sergeant that I had signed on as a medic, and that I had enlisted. So once again I was told to get back in line. I will add here that if you ask any person who had ever served in the military what their service number is, they will tell you without thinking about it twice.

Then we were told we were going to go on police call. Now that sounded cool. There must be a riot somewhere or a student protest, or a drunken brawl. Give me a helmet and a belly club, I'm ready to go! Once again, my ignorance kicked in. I found that going on "police call" meant picking up trash, mainly cigarette butts! What a downer. Welcome to the Army.

This mistake is superseded only by the notion I had that Ft. Jackson was akin to Ft. Apache, on Rin Tin Tin, the TV series. One revelation after the other. Getting off the farm was becoming a real eye-opener for Pvt. Ark! I don't believe there were any TV shows on at the time that showed what a modern-day fort looked like, only westerns. So, that's my excuse.

BOOT CAMP

So, we shouldered our duffel bags *and* were then transported from the Reception Center to an area on Fort Jackson that had the nickname, "Drag Ass Hill." More about that later.

We were then assigned to an old WWII barracks on the lower level. This is where we would sleep, shower, keep our uniforms and footlockers in order, go to the bathroom, etc. This barracks would be my home for eight weeks. In the bathroom area there were a long row of toilets with no stalls or panels separating each toilet. This translates into "no privacy." Not my cup of tea for sure!

Left: Basic Training, Ft. Jackson, 1968.

A few of us were selected to go to the PX to get whatever necessities we needed that we didn't bring from home. At the PX, I picked out shaving crème, Aqua Velva, roll on deodorant, soap on a rope, a double-edged razor, hair shampoo, a Snickers candy bar, peanuts, chewing gum, and a bag of plain M&M's.

Upon returning to the barracks, the sergeant inspected what we all had purchased. I should have hidden my candy on my person, because the sergeant did not care for what I had purchased. He began yelling and tore open my M&M's and threw them around on the barracks floor. Next, were the nuts. He then ordered everyone in the barracks to clean the barracks floor and pick up any candy or nuts that was lying around. What a beginning.

Laying on top of each bunk was a folded wool blanket, a bedsheet, a top sheet, a pillow, and a pillowcase. We were all then instructed and shown how we should make our bunks, every detail had to be attended to and perfect, and once completed, you should be able to drop a quarter on top of your blanket and it should bounce up. Now, that's a tight bunk!

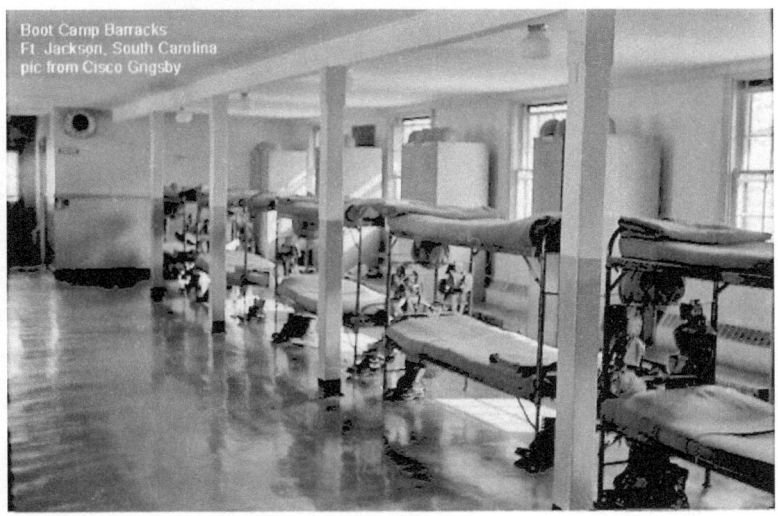

Boot Camp Barracks
Ft. Jackson, South Carolina
pic from Cisco Grigsby

Sgt. Earnest Barnes was my drill instructor for the 8 week cycle I was embarking upon. I liked Sgt. Barnes; he demanded discipline but was not mean spirited like some drill sergeants. He served his time in Vietnam as an airborne infantryman, earning a Purple Heart medal, and he LOVED to run. We ran everywhere. Right across from our WWII barracks was a pop machine with ice cold pop in it. So...guess who was forbidden to use it? Yep, us.

SGT Ernest Barnes, Drill Sergeant

Speaking of running, we always ran in full web gear, sometimes with our M-14 rifles. When my nephew, Jason Hewitt, was in boot camp, just before Desert Storm, gone were the boots and fatigues, replaced with jogging pants and tennis shoes. Dang!

In Basic Training, I became good friends with a guy who played in a very popular rock group at the time called, "The Fifth Dimension." He was the organist for the group and had received his draft notice while playing at Caesar's palace. Talk about a culture shock! When we first met, we hit it off and started talking about where we had come from. I told him that I played a little guitar, and he told me that he played in a band before the Army. I asked him what the name of his band was, and he said, The 5th Dimension. I asked him if they were a pretty good band? Boy, did I reveal my ignorance. The song, "Stone Soul Picnic" had just been released and Howie is playing at the very beginning.

His name was Howard Albert (Howie), and we remained good friends all through eight weeks of basic training at Ft. Jackson and ten weeks of AIT (Advanced Individual Training) at Ft. Sam Houston. Howie was not used to the physical rigors of basic training so if I saw he needed a boost, I'd help if I could.

At the end of most day's activities, we were all usually beat, but there was always one thing left to do, climb up Drag Ass Hill. And sometimes we were asked to do a two man carry up Drag Ass Hill. It was very hard to do at the end of a training day, because our asses were dragging already.

The infamous "Drag Ass Hill."

DRAG-ASS HILL   FT. JACKSON   SOUTH CAROLINA

While in basic training, Howie was receiving 5th Dimension royalty checks every month, which was a substantial amount if I remember correctly, and I received $98 every month, but I wasn't complaining, since the Army had given me all this free stuff.

There was a squad leader in our barracks whose name was Samuel Baker. He was a good soldier and a good person. I will write more about him later, but he will be shot in an ambush in Vietnam three times with an AK-47 rifle.

Samuel Baker was shot three times with an AK-47 in Phouc Vinh, Vietnam.   I was treating wounded   in an aid station when he   was brought in.

A Greenon High School classmate and good friend, Jeff Michael, had started his basic training about 6 weeks ahead of me. We touched base once for a short visit. He headed to Vietnam after combat engineering training. I was Jeff's best man at his wedding. Thankfully, Jeff made it through his tour in Vietnam without a scratch.

Our daily regimen in boot camp usually varied, but it seems like we did a full PT test before every meal, everything but the mile run.

The monkey bars really built up my hand strength but at the same time created many blisters on the palms of my hands. I had acquired big blisters underneath each finger on the palms of my hands. When those blisters became calloused, doing the monkey bars was much easier. Before every meal, we were made to go the length of the monkey bars seven times without dropping.

I lost 30 lbs. in boot camp, weighing 170 lbs. with a 32" waist. Boy, do I long for those days, not boot camp days, the weight.

Aside from the continuous exercise, I typically only ate one meal a day, and that was breakfast. Breakfast was my favorite meal of the day.

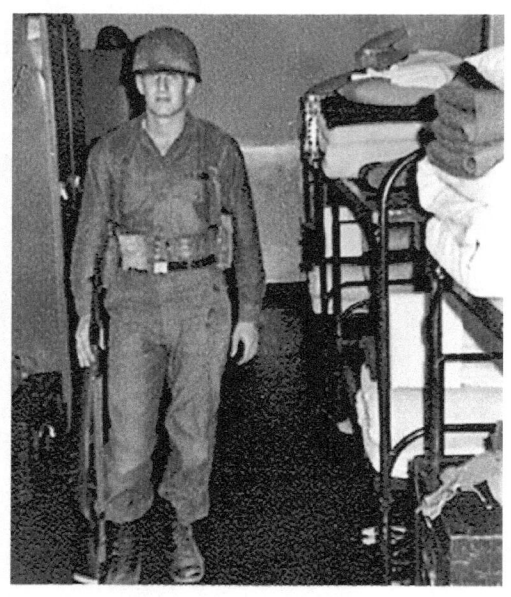

By the end of boot camp, we were all in pretty good shape and our company won the final PT Challenge at the end of our cycle. My father and mother and two of my younger brothers, Bruce and Danny, and my sister Nora came to see me graduate and almost didn't recognize me because of the weight I had lost. It was good for me to see them.

Before I graduated, I was asked if I would like to go to officers training school. Being an officer would require me to stay in the Army another year beyond my three. I decided that I didn't want to do that. In retrospect, as an officer, I probably would not have been a field medic, but then I would never have met Billy Keegan, David Crilly, and Jesse Fugate. And they played a huge part in my life in Vietnam. It's a real eye-opener when you realize that every little decision that you make can change your whole future and destiny. Too bad we don't realize that when we are younger.

FT. SAM HOUSTON, SAN ANTONIO, TEXAS

After completing basic training, Howie and I were flown to San Antonio, Texas to be stationed at Fort Sam Houston for ten weeks of medical training.

When our medical classes began, I paid very close attention because I realized early on how important it was to know what to do as a medical corpsman and that lives might hang in the balance depending on the decisions I made when administering treatment, under fire and otherwise.

Although we were made to watch films of wounded soldiers being treated, and some were gruesome, they didn't come close, at least to me, to the reality of what you see in combat. It is amazing how contorted a body can become due to being wounded or blown apart and how some bodies look with shrapnel and bullet wounds severing limbs, or creating sucking chest wounds, or when intestines are laying outside of wounded bodies and the soldier is still alive. Thinking about it now is sometimes worse than when it actually happened.

We practiced starting IV's, taking blood, giving shots, suturing, stabilizing broken limbs, clearing airways, preventing shock, treating shock, applying tourniquets, clamping off veins and arteries, treating sucking chest wounds, and learning when and when not to administer morphine. I found out that the Army was very strict on the use of morphine and kept close track of how many viles you had in your possession.

I don't remember doing a whole lot of marching or running or even firing a weapon, but I suppose we did. One thing that sticks in my memory is when we were standing in formation one morning and someone in another barracks started playing music loudly out of their window, "Good Day Sunshine," by the Beatles. That made me homesick.

Another thing I recall is that two Led Zepplin songs, "Good Times-Bad Times" came out and also "Whole Lotta Love." And for some reason I remember Archie Bell and the Drells playing "Tighten Up." Go figure.

Howie and I did a lot of things together when we had time off: we went to a few drag races on weekends; a Billy Graham Crusade in San Antonio; and we visited the Alamo, of course, and the Air Force Academy. I remember feeling kind of awesome standing where David Crockett and Jim Bowie once stood and being at the place where they died fighting. Watching Davy Crockett on Walt Disney was a highlight of my young life. That was in 1955. Good times.

It was so very, very hot at Ft. Sam Houston. I remember sweating as soon as I dried off with a towel after a shower. I saw my very first tarantula while on KP duty one morning with Howie. We always had KP together because our last names both began with "A."

One night, Howie was accosted while lying on his bunk, by a guy holding a pocketknife and demanding money. Howie grabbed the perpetrator's knife hand with one arm and his throat with his other hand and squeezed. The guy gave up quickly and left and I don't know what happened after that. Turns out, Howie had a brown belt in karate. Pretty cool, I thought.

During one of our footlocker inspections, a captain was questioning the guy whose bunk was on top of Howie's. He was told to open his footlocker and when he did, we saw that his footlocker was filled to the brim with cartons of cigarettes, very neatly stacked. The captain was not amused. I was, though.

Another guy in our barracks tried to talk me into going AWOL with him and fly on Icelandic Airways to Sweden. "There's free love in Sweden," he says! I told him I'd pass.

We all were given many immunization injections, and the shots were given with an air gun. I assume the shots were to help prevent certain illnesses overseas, but I didn't care for air gun immunizations where the medicine was blown into your arm by an air blast. Needles were less painful.

There was a huge assembly held at the conclusion of our ten-week cycle. I was sitting in this big auditorium talking to some guy during the program and suddenly, he says to me, "I think they just called your name to come up there," and to my surprise, I had been selected as "Trainee of the Cycle," and received a certificate.

Later that same day, I also received orders for Vietnam. "Would somebody please give me a cigarette!"

Right: Pvt. Randall W. Ark, AIT Medical Corpsmen.
July 1968, Ft. Sam Houston, TX

Advanced Infantry Training for Medical Corpsmen.
July 1968, Ft. Sam Houston, TX.

DEPARTMENT OF THE ARMY
HEADQUARTERS U.S.ARMY MEDICAL TRAINING CENTER
FORT SAM HOUSTON, TEXAS 78234

AKPSH-T-T   16 August 1968

SUBJECT: Letter of Commendation

PVT Randall W. Ark, RA 11 838 458
Company D, 4th Battalion
U. S. Army Medical Training Center
Fort Sam Houston, Texas

1. I take this opportunity to commend you for having been selected
as an Outstanding Trainee of Company D, 4th Battalion, USAMEDTC.

2. In a class of 94 trainees, you were selected by your cadre and
peers as the trainee who most distinguished himself by example and
service as the Outstanding Trainee of your class.

3. It is a pleasure to have personnel with your initiative, ability
and interest as members of the Army Medical Service team.

                          CHARLES C. PIXLEY
                          Colonel, Medical Corps
                          Commanding

HEADING HOME FOR 30 DAYS LEAVE

I truly don't remember a whole lot about those 30 days at home before heading to Vietnam. Near the end of my 30 days, some of my Greenon classmates threw a huge going away party one night at Jim Maurer's barn, a place where many classmate parties occurred over the years, and they gifted me a cigarette lighter with my name engraved on it and the date (8-23-68). I still have that lighter, but it no longer works as it once did. My classmates are the best. Shown in the picture are Roger Clem, John King, Jim Maurer, Jon Allison and Roger Maurer.

Left: My friends had a farewell party for me in Jim
Maurer's barn, our major party place.

LEAVING HOME

As I stood in the kitchen of our farmhouse in my khaki uniform, I looked around slowly, thinking that this might be the last time I would ever see this kitchen. I took it all in, wanting to remember every detail. It had to last me a year.

Leaving for Vietnam, August 1968.

Tears filled my mother's eyes as she saw the tears fill my own. She had been through this twice before, once with my father, and then again with her brother, both going to war in WWII. My father sensed my emotional state and quietly told me that I would feel much better once we were on the road and away from the farm. He was right.

Being young and ignorant about the war in Southeast Asia probably worked to my advantage in those days; otherwise, I may not have endured the trip I was about to take. I entered the Army in April of 1968 at the age of nineteen, I turned twenty in basic training and twenty-one in Vietnam. I found out after returning home from Vietnam, that a fellow classmate and friend of mine, Billy Bloomfield, was killed on June 2, 1969, my 21st birthday.

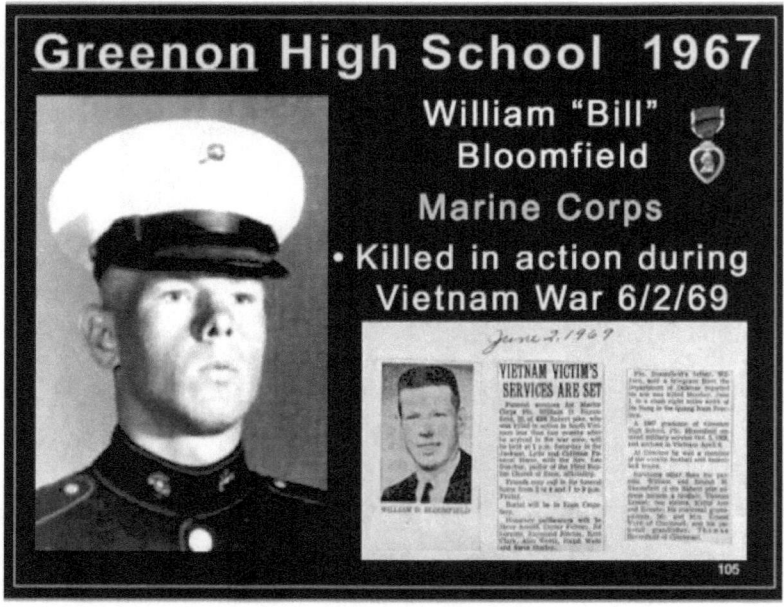

My sister, Nora, my girlfriend, Cheryl England, and my parents took me to Vandalia Airport in Dayton, OH, to see me off. Things were different back then and boarding the plane was not like it is today. Passengers had to walk across the tarmac outside and then up the steps of the aircraft to board the plane. I don't remember much about saying my goodbyes, and I can't imagine what mom and dad were thinking.

I freely admit that there are many events that I am no longer able to recall from this time in my life, and my flight to California is one of them. I do, however, remember landing in Oakland, California and processing in Oakland, around 3:00 a.m., and entering a large hanger with many rows of cots. We were told to pick a spot and wait until our names were called and then we would board our plane to leave our country, our families, and our friends.

I recall a little of the flight to Hawaii, partly because we were only given a hotdog to eat and a coke to drink on the way, not that I was hungry. I remember watching the sunset over the ocean from the plane window, and how surreal and beautiful the sky was. I remember, too, the uneasy feeling of never seeing my home or family again. It was a very melancholy moment, and I was feeling sorry for myself.

After a brief stop in Hawaii, which looked beautiful from what I could see from the airport terminal, and another stop at Clark Field in the Philippines, which was the antithesis of Hawaii, we landed finally at our destination in The Republic of South Vietnam. The airport was located at Bien Hoa (ben wah), South Vietnam.

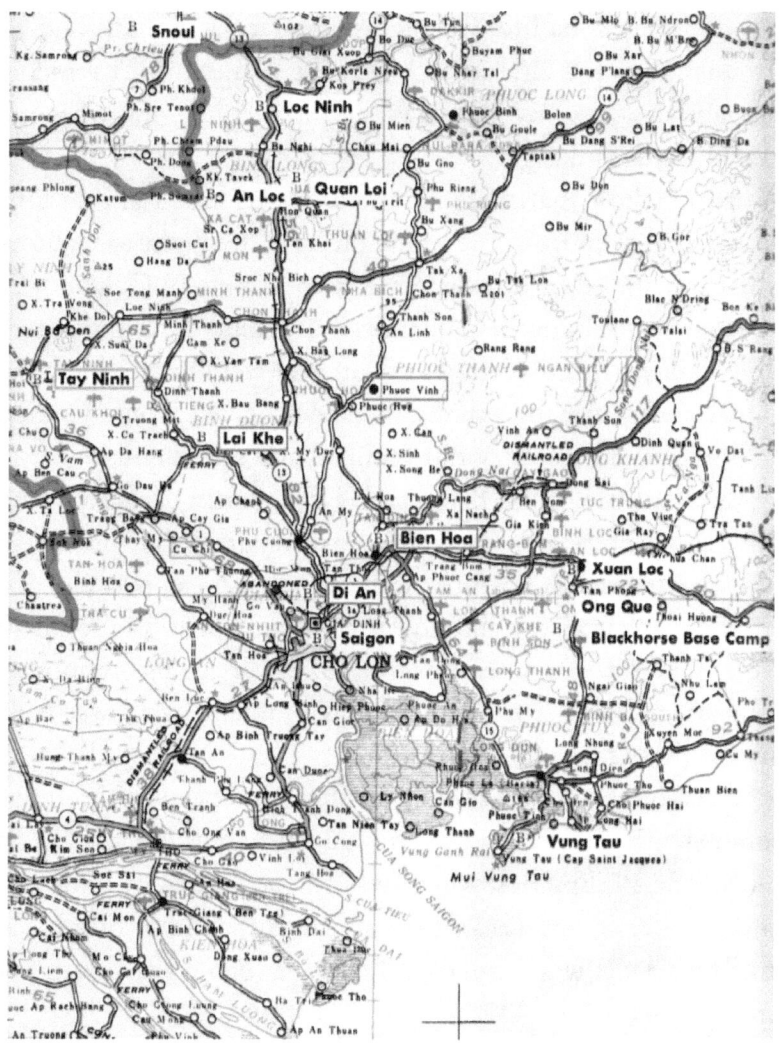

Well, here we go. Stepping off the plane felt like stepping into an oven and the air carried the smell of body sweat, cordite, and burning feces. As we unloaded and walked across the airfield, we headed for soldiers who were waiting in line to go home on the same plane we had just arrived on. They looked a bit disheveled, and some were unshaven. Some had bush hats on, but most had no hats at all. We all kind of expected to be heckled a little bit, since we were new in country and they were leaving, but there was none of that. What I recall the most, however, was the look in their eyes and on their faces. Looks that betrayed that they had seen too much of the war and too often. Boy faces and men faces, with far away looks that begged for peace and a good night's sleep. No one smiled as we walked past these war-hardened veterans, and no one spoke. Oh, there was an occasional, "Good luck!" or "Give 'em hell," or "Watch where you walk," but that was it. Maybe to them, we were dead men walking.

Once we were all unloaded and waiting to be processed and assigned to units, I was hoping that I might see someone familiar from basic training or AIT (Advanced Infantry/Individual Training). I did not.

We then proceeded to a building where assignments were given out. It was there that I learned that I was assigned to the 1st Infantry Division, the 'Bloody Red One." I remember thinking, "Oh no, not the FIRST Infantry Division. They probably see more action than anyone!" Wouldn't they be the first ones to go anywhere, where the action was? I did think the 1st Division patch looked cool, though. I know now that the 1st Infantry and the 29th Infantry Divisions were the first divisions to land on Omaha Beach in WWII.

I don't recall when or where I was issued fatigues, a weapon, or any gear at all, but I am pretty sure that it all happened in a base camp called Di-An, which is pronounced "Zee On". It was also at Di-An that I was assigned to a medical battalion.

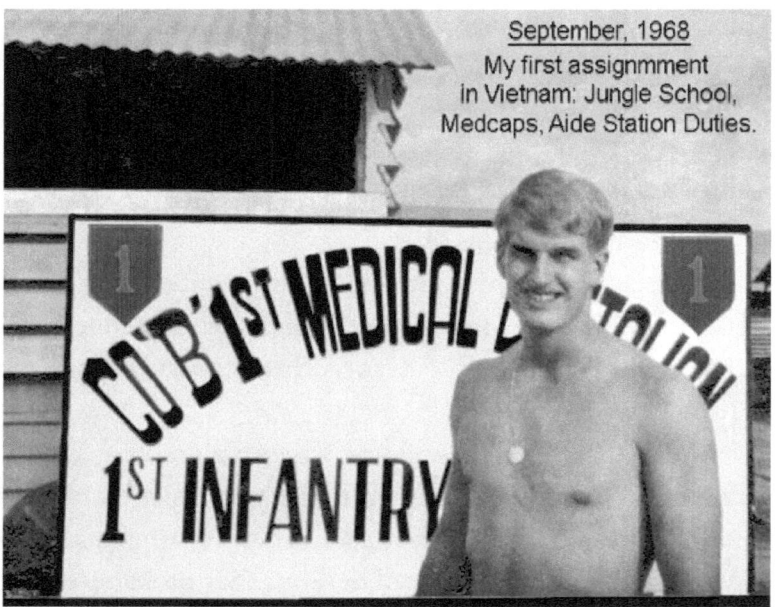

September, 1968
My first assignmment
in Vietnam: Jungle School,
Medcaps, Aide Station Duties.

CO 'B' 1ST MEDICAL BATTALION

1ST INFANTRY

After eight weeks of basic training at Fort Jackson, South Carolina, and ten weeks training to be a medical corpsman at Ft. Sam Houston, San Antonio, Texas, I was now a full-fledged medic in Company 'B', 1st Medical Battalion, 1st Infantry Division, The Republic of South Vietnam, which was a battalion consisting only of medical corpsmen, doctors, dentists, dental assistants, and other related medical personnel.

It was also at Di-An that I had my first war experience that scared me so badly that I shall never forget it. Remember the movie "The Time Machine," based on the book by H.G. Wells? There was a siren used in a future time to call the peaceful Eloi underground to become food for the terrifying Morlocks.

The sound of that siren was identical to the one I heard in Di-An one day. The siren was a signal that put the base camp on "red alert," and everyone was supposed to head for the nearest bunker to escape possible incoming rockets or mortars. I ran like a rabbit with hounds running after me, imagining a rocket was following me like a guided heat seeking missile, and I dove into an underground trench covered with sandbags. I was shaking all over. Everyone in that trench was new in Vietnam and we were all scared and we were all smoking. From the outside, it probably looked like a smoke bomb had gone off in there.

After the "all clear" was given, we came filing out to see the damage. There *was* no damage because it was a practice drill, in case there really had been a "red alert." I nearly had a cardiac arrest from a practice drill!!

I was selected one time, along with another guy, to guard our motor pool area in the base camp through the night. We had two cots, two M-16's, and two clips of ammo. We guarded in shifts throughout the night, one sleeping while the other canvassed the motor pool area.

What I remember most about this particular duty was hearing the eerie sounds of reports of machine guns, M-16's and AK-47's off in the distance. It still gives me chills thinking about it. Somebody, somewhere, was involved a firefight and men were being killed and wounded.

Another thing that I recall were the trucks that drove through the base camp area spraying mosquito repellant everywhere. This repellant, I came to discover, was straight DDT.

And I can't forget a Hawaiian veteran, Jack Mishichima, who taught me bar chords on the guitar. He did that so I could play the background chords to "Light My Fire" by the Doors while he played the lead guitar. Jack was a very nice guy.

After a short time in Di-An, we were shipped by C-7A or 7B Caribou plane to a place called Lai-Khe, also known as "Rocket City."

There were a lot more rubber trees in Lai-Khe than in Di-An. They were big, tall, rubber trees and I was told that our government had to pay for every rubber tree that we destroyed in that country. Go figure.

So, there I was in a medical battalion with a bunch of medics, wondering what was going to happen to me next. Everything was happening so fast. In March, I had been in college, and by September, I was in a war zone in Vietnam. At that point I was beginning to wonder why I didn't try a little harder in college.

Lai-Khe, where I was stationed for the moment, was actually scenic, and it seemed safe enough. I made friends and bought my first 35mm camera from medic Jim Sproul, a Yashica 135 Electra. Jim had only two weeks left in country and was "down-sizing." Jim was a good guy, and I was happy for him that he was going home.

Jim Sproul sold me my first 35mm camera just before
he left for home, a Yashica 135 Electra.

1968
Lai Khe
Vietnam

Simon
Randy Ark
Michael Blaha
Jim Sproul
Bill Keegan

Once we were settled in, all of us new guys were told that we had to go to be trained at a jungle school, and we would be gone for one week. Jungle School is where we would learn about how to operate and use every type of weapon they had over there and to go on night patrols and learn camouflage techniques. I was beginning to wonder when I would be assigned to a nice, air-conditioned hospital, with pretty nurses all around. Ok, I'll be patient.

When first arriving in Vietnam, and after being assigned to the 1st Infantry
Division, everyone had to do a week of "jungle" training. In that week, we
were in the boonies (God only knows where) and we went on patrols and
learned to operate about every weapon they had. We went on day patrols and
night patrols. We learned search and destroy, ambush, and recon techniques.

One weapon that was new to me was the M-16. I had never
held one before, because we were all trained with M-14's in basic
training at Ft. Jackson. As it turned out, I was never to use or
carry an M-16. As a medic, I found it was too cumbersome, so I
chose to carry a .45 pistol. Little did I know at the time that my
pistol would one day become the instrument of death for my good
friend, Pvt. David Crilly.

There is one incident there that remains in my memory. We
were sent out on a night patrol, I am assuming as a practice exercise, and a fellow soldier and I found ourselves separated from the
rest of the group. It was quite dark and we were told to be quiet.
We were neck deep in water between two high ridges, holding our

weapons above our heads when all hell broke loose on the ridges above us. There were red tracers from machine guns flying in one direction, and some other color (probably green) tracers answering from the other direction. We were frozen in fear, but still unnoticed in the water below. We began a slow trek backward out of the area and somehow found our missing patrol.

We all headed back to our camp, only to discover in the light, that our bodies were covered with leaches. Some of the leaches were on the outside of our clothes, and some were on our necks and underneath our uniforms. One guy had a leach covering his eyelid. We used sulfur power and matches to get them off. I was glad when jungle school was over. Only 50 weeks to go!

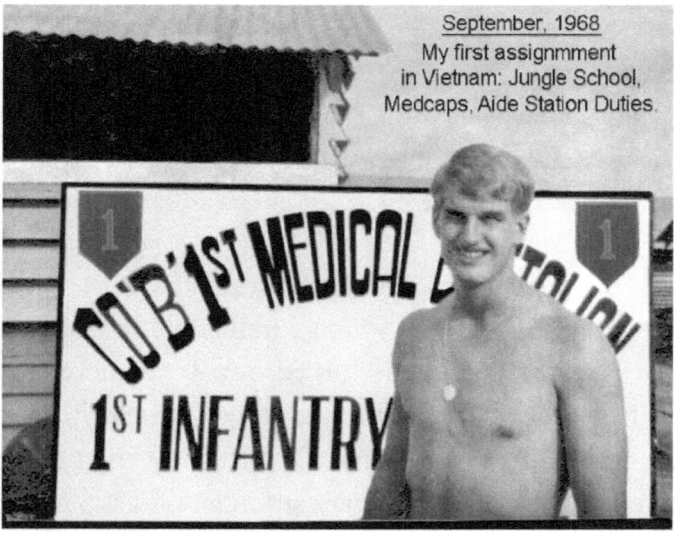

September, 1968
My first assignmment
in Vietnam: Jungle School,
Medcaps, Aide Station Duties.

In our medical battalion, we had all sorts of duties and assignments, some more exciting than others. On Christmas Eve in 1968, I was working alone in the battalion aid station when an MP walked in, dripping in blood. His face was covered with blood, and

I was amazed that he was even able to stand upright. His buddy, who helped him in, told me that they had been in a jeep accident and that the machine gun mount had cut his head open. I helped him onto a table, and yelled for the doctor who was on call. We discovered that it was the gun turret that had sliced his scalp down to the skull in the rear of his head, and that he needed stitches. I now understood how scalping someone was done and the effect of it. I could grab this man's hair and lift his scalp from his skull. The doctor told me that scalp wounds tend to bleed excessively because of the tightly packed blood vessels in that area. I cleaned the wound, shaved the surrounding hair and helped stitch the man up. That is how I spent my Christmas Eve in 1968. It was not a silent night for me.

On New Year's Eve heading into 1969, I was once again, manning the battalion aide station by myself, there at Lai-khe. What a nice surprise it was when fellow medics and dental assistants came to the dispensary to help me bring in the new year! What a great surprise and how thoughtful it was of them to do that.

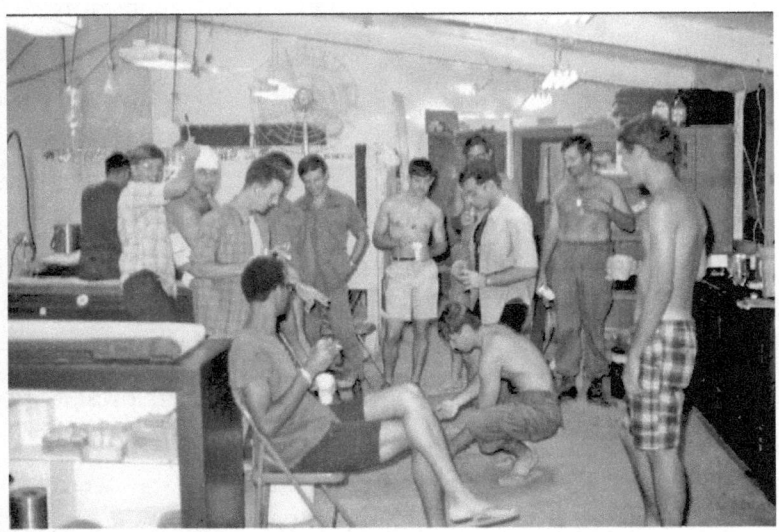

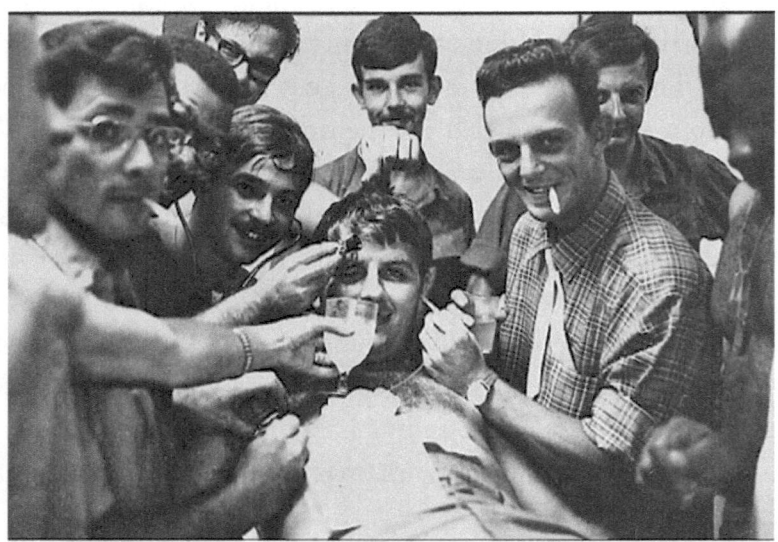

Something I looked forward to in this battalion was going on what we called "medcaps," (Medical Civil Action Programs.) We medics would load up trucks and jeeps with medical supplies, enlist an ARVN interpreter, and head to one of the many area villages in the surrounding area to treat the local inhabitants there. The villagers were usually glad to see us, especially the kids. They always begged for candy and food. An infantry unit would always accompany us for protection and I am glad they did.

Penicillin was a wonder drug in Vietnam and there were certainly many uses for it there. It was sad seeing the little children with open sores and shrapnel wounds, accompanied by flies and infection. There was one little guy I worked on, patching up both his legs. He followed me around the entire day, just watching me assist the doctors and other more experienced medics. I often wonder what became of him.

The villages were very crude compared to the way we lived in America. Some, of course, were better than others, but most were

grass huts or stone houses with tin roofs. I saw children with absolutely nothing to wear, completely naked. It was like going back
in time. There were oxen pulling wooden carts with huge wooden
wheels. It was a simple life to be sure, and at night, there was always
the fear of VC (Vietcong) coming into the villages to plunder anything that we might have left behind for the villagers. Many villagers had known nothing but war their entire lives. I thought about
that a lot. For them, war was just life and that's the way it was. No
radios, no tv, no telephones, no cars, maybe a Honda 50, and they
lived completely under the thumb of the North Vietnamese.

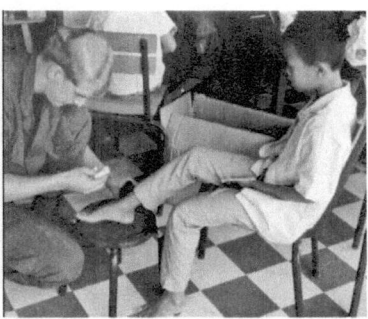

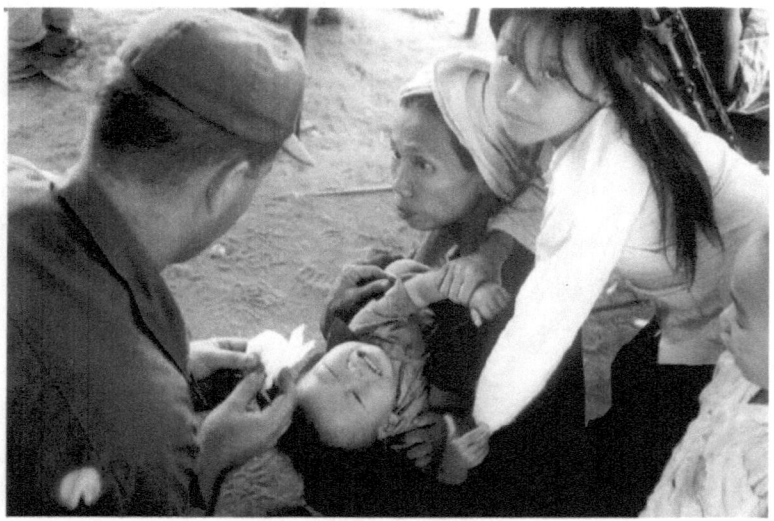

Here is a time when my memory of where we were, or why we were there, is just a haze. I think we were in a little area called Phouc Vinh, and we were helping at what seemed like a M.A.S.H. unit. We were assigned a barracks with cots and blankets for our use while we were there. It was night and we found ourselves in a ward filled with gurneys and cots along with other medical personnel.

Somebody yells, "They're bringing in the stiffs!" Puzzled, I asked someone near to me, "What are the "stiffs?" I soon found out.

A counselor, years later, told me that the dead bodies were called "stiffs" so that we soldiers would be better able to handle the loss and death and not identify ourselves so closely with what was happening.

That we were in a war, and that young men were being killed, hit me like a rock as the bodies poured in. They were in black bags, with toe tags of identification on each one. Boys, my age and younger, lying dead all around the area, alive only an hour ago. I won't even try to describe the smell, save to say that it was the smell of death. I thought of their loved ones back home, thousands of miles away, not yet knowing of their loved one's fate, but I knew. The litters in this ward were soon filled with bodies, some living, some not.

I recalled what war correspondent Ernie Pyle once said, "You feel small in the presence of dead men, and strangely enough, a little ashamed for still being alive. And you don't ask silly questions."

I then saw a bare foot protruding from a black body bag. Inside that body bag was a young man; somebody's son, somebody's brother, somebody's boyfriend or husband. I stood frozen. My stomach felt queasy. I no longer had that youthful feeling of invincibility. I was standing in the midst of my contemporaries, who thought the same as I at one time, that it will always be somebody else.

These are the things I remember. And there are things that most back home can never imagine. Where they might see a list of names engraved into a shiny black granite wall, I would see disfigured bodies lying in pools of blood, bodies blown apart, or saddened eyes realizing their time has come.

You might rightfully ask why I am writing this. Why now? I honestly do not know. I think it's just that sometimes I feel veterans are so alone with their thoughts, thoughts that can never be understood or explained. Thoughts that cannot be shared with just anyone. You may one day see a veteran sitting alone in silence, staring off into the distance with what is called, the thousand yard stare, reliving events that seemingly happened only yesterday. War and combat can do that.

I was told to go assist in an adjacent ward of newly injured soldiers, when I heard someone say my name. "Ark!" It was the voice of a wounded soldier lying face down on a gurney who recognized my voice from basic training. His name was Samuel Baker and he was my squad leader at Ft. Jackson.

Sam had three AK-47 bullets in him. One bullet was in, or near, his spine, and there was one in each leg. I was amazed he was conscious, let alone able to recognize my voice. He asked me what I thought of his wounds, since he knew I was a medic, so I told him there was no blood coming from the holes where he was shot and that was good. He said that he felt nothing from the waist down.

I told him that he would probably be evacuated out soon to a base camp hospital, and I also told him to hold on and try to relax. Tears filled my eyes as I watched him carried to the helicopter. He was my squad leader in basic, and he was a good man. I hope he lived and made it home.

It was- also in Phuoc Vinh that I learned how quickly a soldier could acquire the clap (gonorrhea), an STD. There was a village near to us, and as in most villages, there were a quantity of prostitutes that would provide sexual favors for our young GI's for a small fee, usually $5. The price they really paid would become all too evident upon their return to our barracks. During the night, I would hear screaming and moaning as the Gi's tried to urinate outside our bunker. The burning pain of gonorrhea was too much for these poor souls and they soon realized the risk and consequence of such behaviors. One of the rumors going around was that if a person acquired what was called, "the Black Siph.," you could never make love to anyone again and never go home and there was no treatment. It was a scare tactic for sure, but it worked on some of the guys.

One day we were told that a couple celebrities were in our basecamp if we wanted to go see them. The celebrities were Joey Bishop and Tippi Hedren. If you recall, Tippi Hedren starred in Alfred Hitchcock's movie, "The Birds." It was neat seeing them and I had my picture taken with Joey Bishop. I remember seeing him on Ed Sullivan.

Joey Bishop and Tippi Hedren (The Birds) came
to speak with the soldiers at Lai Khe.

Joey Bishop

It was getting close to Thanksgiving and I wanted to get something for my family at home. There was a little village near our post and I walked to it with some other guys thinking I might find something there. I found some very colorful smoking jackets and they were pretty inexpensive so I bought one for everyone. I asked them to wear them at Thanksgiving and send me a picture. They did.

Thanksgiving 1968

In a medical battalion, it is always in the backs of our minds, the prospect of being sent into the field, which is what we called the front lines. There was one medic we were close to who was the first of our group assigned to a line unit, only to return in two weeks with a psychological disability. He said he just couldn't handle it. That made us a bit nervous.

Bill Keegan was a good friend of mine whom I met for the first time in Lai-Khe. Billy and I became very good friends and remained friends even after coming home. He loved the rock group, "The Young Rascals," and he smoked those god-awful Marlboro cigarettes. Billy was from Pittsburgh, and he and I often talked of being sent to a front-line unit and wondered if we could do it, especially after what our fellow medic experienced. We worried about knowing what to do when the time came that we had to work under fire on a wounded soldier.

I didn't want somebody to die because I forgot how to treat him, or froze from fear, or did not act fast enough. I would lie on my cot at night and think about all the different possible wounds and how I would treat them.

Well, the thing that I had feared most finally came upon me. I was assigned to a 5th Infantry field headquarters battalion aid station near the perimeter of Lai-Khe, and Bill Keegan was shipped to a 1st Cavalry Unit. My life was about to make a radical change.

My buddy, Howard Albert also went to Vietnam as a medic and was wounded by a claymore mine soon after and had to be evacuated to Japan. I visited him there on two different occasions when I had an R & R.

I immediately disliked my new assignment with this 5th Infantry Battalion Aid Station and my attitude reflected that. These soldiers were not representative of an Army of which I wanted to be any part. They did as little as possible, and their primary function seemed to be avoiding anything that resembled helping someone, except themselves.

One day, when our base camp went on red alert, because a rocket had hit somewhere in the area. I just went to my hooch and sat on my bunk awaiting the "all clear." A Captain wandered into the barracks to check out the area and I failed to stand at attention when he entered, and he scolded me about being insubordinate. He also told me that my type of disrespect would land me in the front lines, and he further stated that I could be out of there the next day if I didn't shape up. I told him, "Fine, do it!" He did.

The next morning, I found myself walking into D Battery, 8th Battalion, 6th Artillery. It was to be my home until the end of my tour. I suddenly felt the ground tremble and I was afraid it might

be an earthquake. There was a soldier nearby and he saw my trep-
idation and told me to look toward the horizon. I then saw huge
plumes of smoke in the distance. "That's a B-52 strike, he said, it's
about 20 miles from here." "And you can feel the ground shake
when it's that far away?" Wow.

I was the only medic in this battery that consisted of four howitzer sections. Each section had its own 8" cannon and these cannons could fire a 240-pound projectile a distance of 20 miles. The
8/6th was a mobile artillery line unit based on the perimeter of
Lai-Khe. We were about 150 yards from the jungle and my hopes
of ever being assigned to a air-conditioned base hospital with pretty nurses all around soon faded away.

Lai Khe, Vietnam
8th/6th Arty
1969, 8"Cannons

There were new faces and many new activities that I needed to become used to. I also had a certain duty that most people would cringe at, which I did not mind at all. This particular duty was the burning of human waste. Our latrine had five or six holes and underneath each hole was ½ of a 55 gallon barrel. It was my job to pull out the barrels when needed, mix the contents with fuel oil, stir it all up into a soup, and then burn it. It was all quite neat and sanitary, and it all burned down to nothing but ash. The fun part was when a soldier would come in to do his business and I would stick a hot can underneath him and he would cuss at me, but I'd just laugh and then wait for next victim. Not very long ago, it was determined that the fumes given off from burning the feces were not good to inhale, in fact, a bit toxic. Between that and the mosquito DDT spray, and Agent Orange, my poor body was getting quite a workout.

Battery Latrine

As an aside, it was not uncommon for a Vietnamese woman who worked in our firebase to come into the latrine and sit right next to a soldier while he was "doing his business." I never did get used to that, and usually left when I saw it about to happen. That also reminds me now that I was always apprehensive about being in a shower or latrine with the 24/7 possibility of incoming rounds during such times. Rocket and mortar attacks could happen at any time and at any place.

I had never been around big artillery guns, especially during firing missions. They are unbelievably loud, and the smell of gunpowder was the smell of sulfur. It smelled like a sewer whenever there was a firing mission. A firing mission could come at any time of the night or day, our artillery firing to support troops probably miles away, and often our sleep was interrupted with the booming of cannons. But there were other noises in the night, noises the memory of which still startle me to this day. These noises were the sounds of machine guns in the distance, out in the jungle somewhere, or the sound of incoming rockets or mortars, or both.

Upon arriving in this artillery battery, I was placed into the maintenance bunker and those guys were happy to have a medic handy. My first night there, our battery came under a pretty heavy rocket attack. Naturally, I was nervous and scared. Well, these guys turned on a 16mm projector and began playing a porno video. I have NEVER watched a porno video in my life and didn't intend on starting in a maintenance bunker in Vietnam during a rocket attack. In fact, in my mind, I thought if God is directing where these rockets would hit, this bunker may be a likely candidate. I really didn't think that though.

Maintenance Bunker, a VTR in Maintenance, working on VTR motor.

VTR—Vehicle Tracked Recovery

In a couple of days, I was moved into our battery's communications (commo) bunker so that if there were any wounded anywhere for any reason, I'd be where a radio was located, and I could be sent for.

This communications bunker had two sets of bunk beds in one section and I remember four of us bunking in that area, and it became instinctive to roll out of bed onto the floor at the sound of incoming rounds. We did it in our sleep. There was an exit doorway in this section to the outside and I always felt it would have been too easy to pitch a satchel charge through that doorway and kill us all. I had a top bunk, so my fall was from there, usually onto a fellow soldier already on the ground. There were no complaints though.

From February to the last week of August 1969 this communications bunker on the perimeter of Lai Khe was my home, unless our unit was at a firebase operating from there. Also, it was in this bunker that Davis Crilly died.

In this area were guys who would become close friends. It did not take long in those close living quarters, to know more about each other than even close family members. The guys I remember the most were David Crilly, Jesse Fugate, Raymond Bishop and me.

David was a happy-go-lucky guy who had recently turned 18, so we called him, "Kid Crilly." He was from the San Francisco, California area and came from a large Catholic family. He talked of walking on the beach with his girlfriend and loving the "The Beach Boys." David was fun to be around, and I always enjoyed his company. Getting letters from home or anywhere was a morale boost so I set David up to exchange letters with my sister, Nora, so he would get more mail. My sister enjoyed corresponding with David too, but her heart would soon be broken.

On June 7th, David will accidentally discharge my 1911 .45 pistol over his left eye and terminate his life. I will talk more about this later.

Jesse was from Dayton, Ohio and we looked enough like, we could have been brothers. We talked about playing in bands, "The Diamond Club" in Dayton, guys we hung around with, what we'll do after we get home, and like me, Jesse loved the Beatles and Credence Clearwater Revival. We became very good friends. Jesse died in 1992 from the effects of Agent Orange. My wife and youngest son found his grave on top of a mountain in Hazzard, Kentucky. He was buried in a family plot. Finding out that he had passed away was very hard on me. Here is an article I wrote for the Springfield Paper years ago.

Finding Jesse

On the top of a mountain, off Highway 15 in Hazard, Kentucky, my quest of three years came to an end. It was there I saw for myself, the final resting place of my good friend and fellow soldier, Jesse James Fugate.

I first met Jesse in 1969 at a firebase located on the perimeter of Lai Khe, Vietnam. I was a medic, and Jesse was a radioman. We "lived" together most of the time in what was called, the "commo bunker", and we became good friends. We shared many similar interests, talked of home and we made big plans for "after the war." At this time, we were like brothers, but our plans never materialized.

Jesse contracted yellow jaundice and was shipped to Japan. Even with the best of intentions, after a couple of letters we lost touch. Finally, in September of 1969, I came home, but I didn't think much about Jesse at that point in time.

It was nearly three years ago that I took a strong interest in finding Jesse again, just to touch base and see how he was doing. I made many inquiries and phone calls, used locators on the web, even enlisting the help of friends. My only lead was that I remembered he told me he was from Dayton, but every road I took was a dead end.

A few months ago, I located a phone number in Dayton that was listed on the web as Jesse Fugate. It happened to be his son's number, and I talked with his son's wife. She told me that Jesse Sr., my friend, had died in the early 90's of complications from Agent Orange, and that he was buried somewhere in Kentucky. That was very sad news for me, and it hurt to hear it. I felt badly that he had been dead for so long and I had not known about it. It was information I hadn't counted on or anticipated. She gave me her e-mail address and promised to get in touch with relatives in Kentucky for me. That was the last I heard from her. No phone calls were ever returned, and all the e-mails kicked back. I am not saying it was her fault, it is just what happened.

Now I wanted closure, and I wanted to be sure that what I had heard was correct, though I had no reason to doubt any of it.

As I searched the records of cemeteries and any Fugate's in Kentucky, I happened to remember that Jesse often spoke of an older brother, Curtis. I located a number for a Curtis Fugate in Kentucky, and the age seemed close, so I made a call. Curtis' wife, Wathenia, answered. She was very sweet and most informative, and also promised to get back with me, which she did. She said that Curtis had died recently, and that Jesse had died two years before his mother, in 1992. She said that he had been close to death a few times with health issues after returning from Vietnam

She told me that he was buried on the top of a mountain in Hazard, Kentucky, which is located in Perry County. She later gave me more specific directions on how to find his gravesite. Little did I know what lay ahead in my search for this cemetery.

My wife and our twenty-one year old son, Matthew, indulged me in this quest, and we all headed out for Hazard, Kentucky. To our enjoyment, the drive was beautiful, and Hazard was most pleasant, as were the folks that we met along the way.

After getting a room at the Hampton Inn, we set out to follow the directions given to us by Wathenia Fugate, Curtis' wife. After a few missed turns and a call to Wathenia, we found ourselves at the foot of the mountain where the Fugate Cemetery was supposedly located. We turned off Highway 15 onto a blacktop drive and headed straight up the mountain. The "road" was only suited for one car, and the view from the passenger's window was straight down for 100's of feet. It was at this juncture that my wife and son simultaneously said, "Wrong Turn," which is the name of a scary movie where similar surroundings and circumstance prevailed.

We came to what seemed to be the top, with no cemeteries in sight, but a house appeared with a truck approaching. We asked the driver if he knew of a cemetery up here on the mountain. He said he thought there was one higher up and pointed to a "road" that we passed by on the way up, thinking that certainly, no one would have a cemetery in that direction. We were wrong again!

As we carefully drove back to the where the road took a turn, yet more skyward, we met another man in a truck, who assured us that there were, indeed, three cemeteries on top of that mountain, and that the "Fugate Cemetery" was the last in the line, but all were connected. As I looked up at this "road less traveled", I thought how handy a ski lift would be right now. Up we went, hairpin turns and all.

Behold! The Fugate Cemetery!! My son, Matthew, was the first to find Jesse's tombstone, right next to Jesse's mother's grave.

My feelings were mixed. I was certainly glad that we had found the final resting place of my good friend, Jesse Fugate, and the trek to Hazard was a trip I will certainly remember, but it was again, a reminder to me of yet one more casualty of Vietnam. Nearly 40 years after Jesse and I were there, I have become increasingly more aware of the "total" casualties of that war, and of the varying types of casualties. There were the obvious casualties on the battlefield that I witnessed as a medic. There were post-war casualties, like Jesse, succumbing to the effects of herbicides sprayed liberally on vegetation and humans alike. Lastly, there are casualties that, for lack of a better term, I will call "mental casualties." These casualties are the men and women that continue to live with what they saw and what they did, and then try to make sense of it all, to fit it into their everyday lives, safely hiding the memories from those around them, but haunted day and night by those same memories.

All of this comes to mind, as I stand before the grave of my good friend and fellow soldier. I wish now that I had searched him out years ago, but if I had, I probably would have never remembered him this way and would never have written about him. Jesse was 45 years old when he died. I am glad that I took the time to find him.

—Randall W. Ark

About three in the morning on February 25th, 1969, during my second night in this communications bunker in the 8th and 6th Artillery, we were suddenly heavily bombarded with incoming rockets and mortars. The explosions were very loud and sounded very close. You never know where one is likely to hit. It is very nerve racking.

After about twenty minutes into this barrage, a buck sergeant who reminded me of Wally Cleaver from the "Leave It to Beaver" tv series ran into our commo bunker and yelled, "Where's Doc?" I answered from another part of the bunker and came to Sgt. Leahy. He told me there were wounded men in the first Howitzer section, and maybe more in the other sections. I told him I didn't have a clue as to where these sections were, especially in the dark, and he told me that he would take me there and to follow him. I grabbed my medic bag, and we left the relative safety of our bunker and crouched down at the outside corner of the bunker. We both looked across the open expanse of ground that we had to cover to get to the howitzer sections. We needed to run about a hundred yards across open ground where the only light we had was the light from the explosions of the incoming rockets and mortars. This is what I was trained to do, but I was a bit scared. It was time for the rubber to meet the road.

The sergeant said, "Let's go!" and so he took off and I followed. There was a very bright flash that blinded me for a second, but the sarge grabbed my sleeve and I soon found myself inside the first howitzer section bunker.

The wounded were scattered around on the floor of the bunker, the light was a dim flickering bulb suspended from the ceiling, giving one a feeling of a Hitchcock movie scene. Dust and sand also filled the air.

"Over here, Doc!" The hardest part and the most crucial part of having many wounded men, was deciding who had the worst wounds and treating that person first. Determining this is called field triage'. If a medic erred in this determination, a life could be needlessly lost. At twenty years of age, and only ten weeks of medical training, I was alone in this determination, and the men were depending on me to know what to do. I have thought about this many times over the years, though I try not to.

I asked the men in that bunker to reveal the types of wounds their comrades had. Did they see blood spurting in a rhythmic pattern, or perhaps an open chest wound.

What made matters worse was that these guys were given firing missions and orders for the men to get out on the guns while we were still taking on rockets and mortars. Talk about heroes, those guys were heroes! They needed to help other soldiers' miles away and were not able to help themselves.

With every new mission there were more wounded, not only in this howitzer bunker, but three others. I was running in and out of bunkers, treating the wounded as fast as I could until medical jeeps and evacuation vehicles arrived to take the more seriously wounded away.

If that wasn't enough, our battery began receiving "short rounds," (friendly fire) from a 105-battery located on the other side of Lai-Khe. A short round is an artillery shell landing in our battery area short of where they are supposed to go. Someone quickly got on the phone and located that battery and told them they were firing on us. In those cases, if someone were to get killed, it would be "death by friendly fire." Allegedly, there were over 8,000+ deaths by friendly fire in the Vietnam War.

When the sun came up, and all the wounded were taken to the base camp hospital, I lit up a Kool cigarette and found my way to the commo bunker. On my way there, I heard someone yell, "Thanks, Doc. You done good." Wow, it was great hearing that. That's why I am here, I thought. Forget the politics, I am here to help the wounded and dying. I done good! What a night! I fell asleep.

Upon rising, I discovered something interesting. My fatigues were riddled with tiny burn holes. I removed my shirt and pants. From my neck to my knees I was covered with little bloody red spots that looked more like bloody pimples than anything. I put on better clothes and walked to the base hospital. The doctor there told me that it looked like I was hit with tiny pieces of shrapnel that burned through my fatigues and lodged into my body. He also said that it would be more painful to try to get all the metal out because of their size and that unless all those tiny pieces of shrapnel would be causing me discomfort, they would eventually work their way out on their own. I thanked him, restocked my medic bag, and left. Those wounds were never reported and never written down. It wasn't something I paid much attention to back then; consequently, I was never written up for a Purple Heart. That would happen many years later.

There were many medals handed out for the actions that occurred on the early morning of Feb.25, 1969. If I remember correctly, there were about seventeen wounded soldiers, and many others who voluntarily faced the rockets and mortars to complete several firing missions. There were many Purple Hearts given out and medals for bravery. I received my third Army Commendation Medal, but this one came with a "V" device for heroism. I really didn't know what all that meant until I got home and read the citation that was sent to my parents.

These were some of the men under my care in Lai-Khe, South Vietnam.

Speaking of my parents, my mother told me that one night she heard my younger sister crying and talking in her sleep. She was in her upstairs bedroom at our farmhouse and my mother heard her scream, "No, you can't kill my brother! He isn't dead! You can't kill my brother!" This understandably upset my mother, and she found it hard to get back to sleep.

Knowing the importance of getting mail overseas, my mother wrote to me nearly every day. It took about thirteen days for mail to reach us, and I often thought about relatives back home continuing to get mail from their beloved soldier overseas, after previously getting a notice of his or her death in battle. Mail was our lifeline to home. It was hugely important to us! Howard Albert's mother wrote to me often, though we never met, and she would send me chocolate chip cookies.

One time our battery was sent out to set up a firebase somewhere up north and a classmate and his wife, Jon and Bonnie Allison, sent me a vacuumed sealed ham. I think it was around Easter time. They also sent along a squeeze bottle of French's Mustard. Well, I *had* to share it, so I opened the vacuum tin and then open the mustard. We each took a bite of the ham with our fingers and then squirted some mustard in our mouths. Crude, but delicious.

A lot of the guys asked me to write love letters for them, mostly to their girlfriends or wives or both. They liked the way I put words together, and often they would get some very encouraging letters in response, which would thrill them to no end. I didn't mind at all, and I made some good friends by doing that. I wrote a lot of love letters for a lot of guys to a lot of women. I now wonder how many women received love letters from Randy Ark.

Not all the letters from home were good ones, though. Some men received the dreaded "Dear John letter." One guy received a letter from his wife informing him that she was selling their house and taking their three kids and moving somewhere else, and not telling him where. His next letter contained divorce papers that he was requested to sign. It was very sad, and I felt sorry for him. He seemed like a nice guy.

Another soldier, working on his third tour over there, found out that his mother had spent all the money he had been sending home for the past two and one-half years. He was very distraught upon learning this. Who could blame him?

Speaking of money, we soldiers were paid in MPC's, Military Payment Certificates, I was paid $250 a month as a Pfc. I extra money was alloted for being overseas and fighting in a war. I sent $200 home each month for mom to deposit for me and saved $50 for myself.

On June 7th, 1969, Jesse heard a shot from inside our bunker. He ran and found Pfc. David Crilly with a bullet hole in his forehead just above his left eye, and my .45 pistol lying on the floor. Jesse told me later that blood came out of his head like a faucet. He stuffed a pillow over the wound, and he and some others took Crilly to the base hospital, where he died soon after.

Jesse's eyes were red and watery as he told me about the incident. I will never forget the look on his face. I just stood there silently, trying to take it all in. Crilly was just a kid. I had to write home and tell my sister, who had exchanged a few letters with David. She was understandably very shaken up and the cost of war had once more penetrated our lives.

Of course, I was investigated by the CID (Central Intelligence Division), because my prints were all over the weapon, and it was supposed to be unloaded. It was loaded because the previous night we were on "sapper" alert, and I was left alone in the bunker and told to shoot anything that came through the door unannounced.

A "sapper" attack is when the VC cross the perimeter with explosive satchels and throw them into bunkers and underneath guns. I sat on the floor in the corner of our commo bunker, in the dark, my knees pulled up to my chest, with my arms resting on my knees with my gun pointing toward the entrance. It was too quiet. Suddenly a German Shepard dog come running into the bunker and about scared the daylights out of me. I swore at it and threatened to shoot it, but I didn't.

I had not remembered to unload my weapon. David Crilly was dead at 18 years of age. In case you are wondering, I do not think David killed himself, I think he was just messing around with my pistol and accidentally shot himself. I have always wondered what the Army told his mother. I was to find that out years later from David's mother that she was told that he was riding in the back of a deuce and ½ and his M-16 accidently discharged. She was so relieved when I told her the truth when my wife and daughter drove to California to visit her and David's siblings. We were treated like family and fed a delicious meal. David's brother, Donny, took us to see his grave in a Catholic Cemetery about 50 miles from their house. We located David's grave and cleaned it off, pulled some weeds, placed flowers and flags on his grave.

David Anthony Crilly
Corporal
D BTRY, 8TH BN, 6TH ARTILLERY, 1ST INF DIV, USARV
Army of the United States
Fresno, California
September 29, 1950 to June 07, 1969
DAVID A CRILLY is on the Wall at Panel W23, Line 98
See the full profile or name rubbing for David Crilly

David Crilly fatally wounded himself accidentally
with my .45 caliber pistol, on June 7, 1969.

It wasn't too long after that, that Jesse contracted yellow jaundice and was sent to Japan. I was glad for him and sad for me. I missed my buddies. I discovered in 2007 that Jesse had died in 1992 from complications from Agent Orange. That was so very sad to find out.

I remember that in the early morning of Mother's Day, 1969, we were under rocket and mortar attack for seven straight hours. Think about that; seven hours of rockets and mortars, waiting helplessly, wondering where the next one might hit, waiting for the sound of, "Medic!!" I was scared that night, and I prayed and prayed. The next day I wrote on a Mother's Day card, "Dear Mom, I am in a safe area so don't worry about me. Nothing much happening here." I am sure that my father didn't believe that for a second.

The monsoon season was now upon us, and there are three things that I recall concerning the monsoon rains. At times it rained so hard that it actually hurt to stand in it. Hard to believe, but it's true. I remember that our bunker leaked so badly in one place that one of the guys took advantage of it and had an indoor shower. I took a picture.

Our Bunker leaked so bad, you could shower in it during the monsoons.

Lastly, I remember that I always felt safer when it rained hard, because a sergeant once told me that the VC or NVA would not fire rockets or mortars during a hard rain because the friction of the rain would cause early detonation. I don't know if that is true or not, but I always felt safer anyway, and still get that same feeling, even to this day, whenever it rains hard.

I took two R & Rs to Japan in 1969. R & R means *rest* and *relaxation*. I went to Japan for the sole purpose of seeing my buddy, Howie Albert, a fellow medic who was wounded and sent there. I stayed at a place called Camp Zama and I will never forget my first taste of real food. It was a BLT, with fries, and a large orange juice. Man, did that taste good, especially after what I had been eating for months.

I called home even though I had to call collect. The longer we talked, the more it cost but it was nice to hear their voices. I found out somehow that my phone call cost my parents $50. I decided I didn't want to do that to them again, so I didn't call home on my second R&R, and my mom did not like that at all. I can't say that I would have wanted easy communication like you'd have with iPhones. Communicating with folks and friends at home anytime you want does not seem right when fighting in a war.

I hooked up with Howie and stayed with him for the week. It was hard to get used to walking on the street and being heads above 95% of the people. Howie took me to a back alley eating place one evening, up a stairway in the alley. It looked kind of scary inside, but the food was very good.

Howie also took me to a place where I could buy a decent stereo for a decent price and have it sent home. Howie helped me pick out the components that I needed to create a good stereo system. We bought a Kenwood receiver, an Akai reel to reel tape player, an equalizer, a "zero tracking error" turntable, and Pioneer floor speakers. I had my mother send me the money I had saved up so I could buy these stereo components. Howie was a good friend.

It was at this time that I had to make a life-changing decision. Howie was working with doctors in a surgical hospital, and those doctors, he informed me, were very much against the war. He said that he had talked with these doctors about me and Howie told me that if I wanted, they would fix my medical records so I would not have to go back to Vietnam if I didn't want to. He said they would make up some medical reasons why I would have to stay in Japan.

Well, you can call me stupid, but I felt I had an obligation to serve out my time, and to get back to the guys who trusted in me to take care of them if they were wounded or sick. I also thought about my father, and what he would think.

I thought about Dad often throughout my army experience, and I did not want to be a disappointment to him, or an embarrassment. I looked up to my father for many reasons, and in many ways. I will never forget a letter that I received from Dad once. He wrote that he wanted me to never to feel alone over here in Vietnam, because a part of him was always with me. The letter brought tears to my eyes, and I missed him horribly at that moment. Only *he* could relate to what I was going through and experiencing. Dad would later become my lifeline when I returned home from the war.

We had a laundry of sorts in our area and the
proprietors were MaMa Son and Peanut.

Peanut confided in me one time that the VC had come into her village the night before and she was watching a baby son and she told him to, "Run, Baby Son, Run!" And he ran to hide somewhere away from the VC. For some reason her telling me that affected me and made me think about the way these people had to live.

One time I was walking to the base camp hospital along a gravel road, needing to restock my medic bag, when about 75 yards behind me a rocket hits and explodes right where I had just walked. I turned around to see where it hit and then continued walking. Not long after that, a jeep pulls up behind me with two officers in it. They told me to get down in the ditch until I get the "all clear." I told them there was no need, there won't be any more rockets, the gooks just want to put the base on red alert. "Get into that ditch, soldier!" I got into the ditch, they left, I got up and continued on my way.

Another time I was asked to ride shotgun on a deuce and ½ going to pick up new troops somewhere and take them to their new duty station. As the men were loading up, a captain came up to me and told me to get in the back with the others and that he was now riding shotgun. I said, I can't do that, sir. "That's an order, I outrank you!" I said, "Sir, you are confusing rank with authority. I have the authority here. Please get in the back with the others."

There are a couple things I want to mention that just popped into my head and one was when we took off from Lai-Khe one time to set up a firebase somewhere. Once we had built a bunker to sleep in, I rolled out my sleeping bag, and there was about a 10" scorpion that was in there. I pulled out my Bowie knife and stabbed it and stuck it into the ground. That way, I'd always know where it was. Oh yes, my Bowie Knife.

There was a soldier friend of mine whose name was William
DeShazer, and he was the owner of this Bowie Knife. Well, he and
I were up on top of one of our bunkers checking out his AK-47

when a sergeant came and told him that the captain wanted his AK-47 and Deshazer said he wasn't giving it up, it was his. The sergeant made a move to come up the ladder to get the Ak and Deshazer shot a few rounds at his feet to keep him down. Well, you can guess how that went over. The AK was confiscated and Deshazer was sent to LBJ (Long Bien Jail). Before he left, he gave me his Bowie Knife to hold for him. He never came back. Maybe he'll read this book and come get his knife.

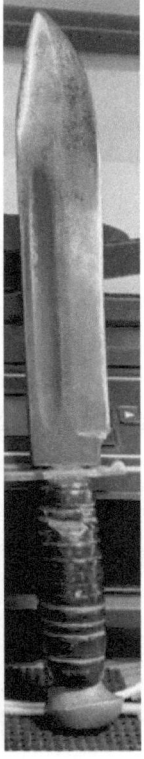

Well, my time was finally growing short in Vietnam. There are so many things that I cannot remember from those days. I know there were many rocket and mortar attacks, a few personal fights, sounds of machine gun fire in the night off in the distance, feeling the earth shake from B-52 bombings 20 miles away, and the constant fear of attack in the night. I remember the 4th of July, 1969, when a light show was put on by Huey helicopters shooting mini-guns just outside of our perimeter. A steady stream of red tracers is quite a sight. There were other displays that the helicopter's treated us to and it was really cool that they did that.

I remember sometimes thinking about what was going on at home while I was in Vietnam. The days and nights were switched, but I often wondered if my best buddies were out on dates at the drive-in, or eating a Tasty Burger at the Fairborn Hasty Tasty, while I was anticipating a night attack. I wondered if anyone back home might be looking at the same moon as I, at the same time? And I will always remember the rats.

Along with the ordinary and accustomed fears that I experienced nightly at this firebase, were the visits onto my bunk each night by the neighboring rats. We all had heard tales of huge rats farther up north that would rip a man's throat out in his sleep, but fortunately, these rats were not so vicious or huge. They lived in the sandbags above our heads during the day, and would police our bunkers at night, scavenging through anyone's leftovers, looking for any morsels of food carelessly left open or dropped to the floor.

My bunk was closest to the ceiling sandbags and the obvious top rung of the "rat ladder" down to the floor. On many a night I was awakened by five to seven rats traversing my bunk. I could feel the tiny depressions of their rat bodies and movements on my thin wool army blanket, and I would awaken and shake my blanket and their rat bodies would thump to the floor or they would scurry back up into the sandbags.

On nights when I did not use a blanket, I would occasionally feel little nibbles of rat teeth on various parts of my body, usually my fingertips, my elbows, or my earlobes. Only once did the little critters break my skin.

I had one week left in country, and I awoke to find teeth puncture marks on my right forearm. I was afraid that I would have to get rabies shots, which is a two-week process, and that I would have to stay in country one week longer than required. I did not relish that thought at all.

I took a jeep to the base hospital to see a doctor. He told me that fortunately for me, rabies were not a big deal over there and that I could get a tetanus shot and that would be it. Thank God. Yes, thank God!

On July 16th, a few of the guys in our bunker and myself were able to watch the Apollo 11 Mission, along with millions of other people, on a very small black and white Sony tv that was in our bunker. I have no idea where that little tv came from, but I was glad it was there right then. I remember astronaut Neil Armstrong saying as he stepped onto the surface of the moon, "That's one small step for (a) man, one giant leap for mankind." How cool it was to be watching this epic event in a bunker in Vietnam.

In Vietnam, when a GI's remaining tour was under 100 days, you were called a "two-digit midget." If you had only one week left, you were called a "one-week wonder." When it came to be my turn, when I had only one week left in country, I became very nervous and very excited. I did not want to be killed in my last week! Had it been a year already? What was I saying? This was the longest year of my life!

At the end of my last week, I bid farewell to my comrades in arms and was taken to either Di-An or Long Bien to process out of the country. It was somewhere in those last two days that I met up with my old friend, Bill Keegan, the fellow medic who was sent to a 1st Cavalry unit. We rejoiced in still being alive and we shared stories of triumph and sorrow, as only combat soldiers can. It was so good to see him again. Billy was wounded also and was put in for a Silver Star, the third highest medal a soldier can get..

At the last stop before boarding the plane home, we were all herded into an area with large circular containers/barrels placed at the front. We were seated and quieted. A master sergeant stepped onto a riser and spoke. He told us about making it to this point in our "tour" and what a shame it would be if any of us had to be kept behind for any reason. We all looked at each other bewildered and wondered how that might happen. He told us about many GI's who had tried to take home black- market souvenirs or drugs of any sort. "Nam grass" was a very hot item in the states, I was told. The sergeant also told us that this was our last opportunity to rid ourselves of any possessions that might possibly detain us in Vietnam. If we would come forward and deposit our "contraband" into the barrels up front, nothing more would be said about it, and we would be free to leave. I guess that a "human wave" would best describe what I witnessed next. It still makes me laugh to think about it. I saw bayonets, AK-47's, .45 pistols, kilos of grass, dismantled M-16's, you name it, being tossed into the large bins. These guys wanted to get home!! Speaking of Nam Grass, I used a kilo of Nam Grass one time for a pillow at some firebase we had set up and another time I used a brick of C-4 for a pillow.

The barrels filled quickly, and we were on our way. As we left that area and stood in line on the tarmac, we saw the new soldiers getting off the plane, new arrivals in Vietnam, and they stared at us as I had once stared at others upon my arrival.

My tour had now come full circle, and I was on the other side of the fence. I had to wonder, as the new guys walked by and looked at us, if we had the war torn looks of the men that we saw when we had first arrived in country one year earlier. Did we have that "thousand-yard stare," we so often heard about with soldiers who had experienced combat and lived to tell about it. I wondered and felt sorry for them. Nothing was said.

Billy Keegan and I were the last two soldiers to board that plane home. What a thrill it was to be on the plane and to feel it taxi onto the runway. As we picked up speed, we all felt it as the plane lifted off the ground. A cheer arose that drowned out the engine noise. We had left the country of Vietnam and were heading home! Home. We made it, we survived Vietnam. What will it be like coming back home? Little did we realize at the time, how much of Vietnam we were taking home with us.

After a short stop in Japan, we headed for Oakland, California. What awaited us there at Oakland Airport is hard to describe. Our welcome home was not what I had anticipated, nor would wish on anyone. There were multitudes of protestors waiting to shout at us, spit at us, and call us names. We heard words like, "Suckers!" "Baby Killers!" and other things not fit for print. Can you imagine protestors curing at us, belittling us. We were coming home from a war, for many of us, it was a year in combat, and to be treated like this. I absolutely could not believe what I was seeing and hearing. How could this behavior possibly be tolerated? How can people belittle combat troops coming home after what they experienced?

I have heard the tales of other soldiers at other times, recalling their experiences coming home and afterward. One Marine had his leg urinated on while standing at attention as an honor guard at an Arlington Cemetery funeral. This particular Marine had been stabbed multiple times during the course of taking a hill close to the DMZ in Vietnam. Another soldier told me of falling asleep in an airport, having only served in the Philippines, and when he awoke, he discovered that he was covered with saliva and phlegm from spit, and chewing gum stuck to his clothing, from people passing by. I could not believe that people could do this to soldiers returning from war or just serving their country. Just forty-eight hours earlier, I was in Lai-Khe! No one had prepared me for this. We made our way through the hecklers, onto a bus that was shielded with metal plates that covered the windows. The bus carried us to a transfer point where we would get a new "dress green" uniform to wear home.

I was so proud of my shiny CMB, (combat medic badge) that I wore above my ribbons over my left pocket. That really meant something to me. I was told that my Army Commendation Medal with a "V" device was a good medal, but my CMB was "it" for me.

Then we received more homecoming advice. The commander was very serious when he addressed us and told us to listen carefully. "Be advised," he said. "When you return home, it will benefit you *not* to wear your uniform in public, do not tell anyone that you served in Vietnam, and do not put anything about Vietnam on a job application or they will not hire you." I was crushed. I was proud to have served my country! What about all those men who had been wounded or died fighting for their country? Did they deserve this treatment? I didn't know what to think or say.

Bill Keegan and I were still together, and we would not part company until we hit Chicago. While in the Chicago Airport, a man in civilian clothes walked up to me a said, "You need to button up your coat there, soldier." I said, "Who the hell are you?" Turns out he was a Full Bird Army Colonel. I buttoned up my coat. Keegan and I vowed to keep in touch and so we did. Bill passed away in 2023, just after my family's visit with his family in Pennsylvania. I will surely miss Billy!

I arrived at the Dayton-Vandalia Airport on a Sunday morning. I had pre-arranged for Cheryl England, the girl I had been dating before leaving for the service, to pick me up and take me to my family's farm from the airport. I would have never found my way home otherwise, there were so many new roads and overpasses. I don't remember much about our drive home and I asked Cheryl about it years later. She said that I was very quiet and mostly stared out the car window.

We finally reached our destination, our family farm on Rebert Pike. Cheryl drove me down our farm's lane, and I asked her to leave me there alone, and that I would contact her later. I thanked her for picking me up, grabbed my duffel bag and headed into the farmhouse. Years later, I asked Cheryl how I acted on the ride home after she picked me up and she told me I was very quiet and stared out the window mostly.

Everyone was at church, except my brother, Bruce, and he was surprised to see me, as I had told no one that I was coming home. I just stood and looked around the living room, not knowing really what to say. Then I heard a car pull up. It was Dad. I stepped out onto the front porch as he got out of his car. He saw me and stopped. He walked slowly toward me, wrapped his arms around me and said, "Welcome home, son." I have tears in my eyes as I write this. I can still feel the warmth of his love and the depth of his caring as we embraced.

My mother and sister were the next to arrive from church and I could hear my sister Nora screaming before the car stopped. She ran and jumped into my arms and said, "I am so glad that you are home!" More tears. I am assuming that my youngest brother, Danny, was with them, but I do not recall. I also do not recall my older brother, Steve, or my younger brother, Nick. I am not saying they weren't there. It is just that I can't recall.

My first few days at home were very restless. I could not sleep; it was too quiet. I did not drink or do drugs in Vietnam, because I wanted to have my wits about me if we were attacked, but I soon took up the habit of downing a pint of Southern Comfort each night before I went to sleep. I don't know if my parents were aware of this or not, but they never said anything. I sat alone in my bedroom on the farm, and in the darkness, sipped my way into sleep.

I was alone in my thoughts, completely alone. I was unsettled and very mixed up about my thinking patterns and my psychological well-being.

My father later confided in me that he thought I had come home too quickly from Vietnam. He thought that the transition from battlefield to civilian life was too quick, and that most WWII vets had had time to process things a bit, and in many cases, had weeks to talk with fellow soldiers while waiting to leave Europe or wherever, and on the way home, aboard ship.

There were classmates who wanted to have me over for a meal and they would ask me what I would like to have, and they offered steak, roast beef, etc. I would just tell them hamburgers or hotdogs, baked beans, cole slaw, bacon-lettuce-tomato sandwich and chips. Really, they'd say. Yep, really. Keep it simple, nothing fancy. I'm very low maintenance.

Classmates Roger Clem, Jon Allison and I drove to Niagara Falls just for the heck of it. It was a very nice treat for me and it was something that they wanted to do for me. We rode on the boat called, "Maid of the Mist" Waterfalls Tour and afterward saw the sights. It was a fun relaxing trip that I will never forget.

I was excited looking forward to opening up the stereo components that Howie picked out for me in Japan and I had sent home. I remember setting it all up and I went shopping at Wren's located in the old Bushnell Building on Main Street in Springfield. I began stocking up on every Beatle Album I could get my hands on. And other groups, too.

My Aunt Helen (my father's sister) and my Uncle Johnny

threw a homecoming lunch for me, and my whole family and all my first cousins were there. It was a super time. Aunt Helen's son, Johnny Robert, (aka Bobby) told me recently that his mother, my Aunt Helen, had really worried over me being over in Vietnam. I hadn't realized that.

Bobby was instrumental in helping me get a three month early out of the Army to go to college. He signed me up and registered me at Wright State University and created a schedule for me.

At first, I was thinking, with my medical background which I did enjoy, such as it was, that I might try being a Physician's Assistant. Back then it was a new discipline, and there was only one school available to be educated for that and it was located in the state of Oregon. But, I wanted to go home.

Time to leave for Fort Carson, a new place with new people to meet and new duties.My friend, Bill Keegan, happened to be assigned to Ft. Carson also and that was a good thing for me. It seems like me and Billy were destined to be friends in this life. We both lived in the barracks for a time and Billy had a large poster above his bed that was the front cover of the Beatles Abbey Road Album.

I lived in a couple different places other than the fort, but it seemed like I was always hard up for money and hungry. I did go with a girl named Pat for a while, who was a schoolteacher in Colorado Springs, which is where Ft. Carson is located and we traveled to Mexico with another couple. We drove 250 miles across the border, down to Chihuahua, to see the sights. Wouldn't do that today. Billy and I also lived together for a time with a bouncer at the Old Corral.

I lived with a fellow named Rick Ulrich for a time. Rick and I had become good friends in the barracks and really hit it off being together. He was from Chicago and he knew everything about cars, especially Chevrolets. He got jumped once while walking to our apartment by three guys. He beat them up. He was a tough little guy.

Another friend, Gary Tanner, from California, was into smoking marijuana and I rode with him and his brother in a Volkswagon Beetle to Provo, Utah. They dropped me off at some Mormon Bar while they had some business to conduct elsewhere. This bar had no alcohol, no cigarettes, and I had a long wait.

Billy Keegan and I became close friends with Judy Albin and her daughter Kammi and I am still in touch with Kammi on Facebook. She is a good person with a good heart. I would meet Judy at places like The Old Corral Saloon and we'd dance. We were close friends but our relationship never went beyond friendship.

I worked in the Fort Carson mailroom for a while and the best part about that was taking a jeep and driving up into the Rocky Mountains to deliver mail to soldiers bivouacking. That was cool. I loved doing that.

I began taking courses in criminal law at the University of Colorado extension campus and I was doing quite well and was ready for the first big exam. I was ordered, however, to go up into the mountains with our unit and bivouac. I was supposed to be exempt from that sort of thing if taking classes off the fort. I was threatened with an Article-15 which is not a good thing.

The bivouac area in the Rockies was covered in snow and I have never been so cold in my life. I was attached to a company I was not familiar with, so I didn't know anyone. I tried to sleep in the ambulance that was there, but it was not heated, and it was very uncomfortable with four of us packed in there. I decided to try my luck out on the ground somewhere, but everything was covered in snow. I resigned myself to sleeping in the snow and spread out my sleeping bag on top of the snow. I kept my boots on which I think you're not supposed to do and crawled into my sleeping bag. I truly felt like I would freeze to death in the night. My teeth were chattering and my body was shaking, Like I said, I had never been this cold in my life.

Dawn finally arrived and I was so glad to feel a little bit of warmth from the sun. Well, I peeked my head out from inside my sleeping bag and what do I see? I watched a Second Lieutenant come out of his tent in a white t-shirt and stretch his arms and take a deep breath. "What the heck, I say to myself!" Little did this Lieutenant know how close he came to waking up to a frozen corpse not 50 feet from his tent. I know, I'm getting a little dramatic. Let me just say, it was cold, and I remember it vividly.

The "Cave of the Winds" and the "Garden of the Gods" were very nice places to go. At Ft. Carson, there was mainly one of two places I needed to be during the day, the motor pool or the clinic, to draw blood or to administer injections. Some days I would report to the clinic and tell them I had duty at the motor pool and then tell the motor pool that I had duty at the clinic and then I would head for the Garden of the Gods and drink beer with some friends.

One time, I, and some other veterans went off the post to the

Garden of the Gods at night in the winter and climbed some of the higher rocks there. As cold as it was, I became very warm after climbing a while.

There were a lot of drugs available in Colorado Springs at that time in 1969 and 1970 and the Garden of the Gods became like a marketplace for drugs, mostly acid, weed, or mescalin. You could hike through the trails and voices would come out of the bushes advertising their wares. "Green Pentagon Over Here!" Black Beauty's Over Here."

I spent the remainder of my military time at Ft. Carson, attached to the 4th Infantry, and then later to the 5th Infantry Divisions. I do not recall ever discussing Vietnam with anyone, nor did I have a desire to. My experiences in Vietnam were buried, and I lived for a time as if it had never happened. It would not be until later in my life that the war would once again rear its ugly head and invade my sleep as well as my waking hours.

My older brother, Steve, and his girlfriend, Meg, drove out to Colorado to see me. I remember that he was driving a convertible and that he had a quantity of cans of motor oil in his trunk because his car leaked oil so bad, he needed to have some handy. Funny.

We went for a drive up into Cheyenne Mountain one morning to see the Broadmoor Hotel and to look down upon the city of Colorado Springs. It was a beautiful drive, but in the afternoon a snow shower came through and snowed us in at Steve's Hotel. That's the way the weather is around mountains, it can change very suddenly.

My cousin, Paula Boyer called me and took me to an ice hockey meet at the Broadmoor. Another classmate, Hobey Jay Allen, came to Colorado Springs and we drove to Boulder, Colorado to look around.

And even though I practically lived at the base of Pike's Peak, I never had an interest in going up to experience it. For some reason I had no desire to do that, and neither did anyone I ran around with.

After two years and nine months, the time had come for me to ETS (Expiration Term of Service) and leave the Army behind. Ft. Carson did not make it easy for me to leave. I was made to get three haircuts before I was allowed to leave the fort. Three haircuts! What the heck!

I had a friend whose name was Ronnie, a skinny little blond-haired guy who rode with me as far as Indianapolis where I dropped him off at his home. He was a good kid, and I hope he had a good life. On January 4th, 1971, I was officially out of the active-duty Army. My military obligation was finished.

* * * * *

Arriving home felt good, but I missed being around the mountains. I always enjoyed looking in the distance and seeing mountains. Ohio seemed so flat. That was something I really missed.

After attempting to adjust being out of the Army and home again, I found myself ill-at-ease and was restless.

Since there was no one else in my life with whom I could turn to that could help me understand what I was going through, I turned to my father who had experienced WWII. I remember sitting in our living room at the farmhouse, trying to explain my sleepless nights, and being afraid to sleep, for fear of what I might see. I told Dad that I was jumping at every loud noise and crying for no reason. My father assured me that even though there are some things I experienced that will never be forgotten, they will lessen in intensity.

Things I had seen or experienced were invading my thoughts,

and I had to fight to keep them away. I do not recall any of my friends or family ever asking me about what I went through in Vietnam. I have always thought, maybe erroneously, that if I had had a brother or a friend who had experienced war for a year, that I would be curious about what he did, what he saw, and how he lived, how he felt.

Perhaps they were being courteous and cautious, not wanting to disturb my civilian life by bringing up the war. I don't know. And I also don't know what a good therapy would have been for me at that time. I did not entertain any thoughts of wanting to talk to any other Vietnam vets, and even avoided opportunities. I did not watch any Vietnam movies.

I began college again in 1971 at Wright State University, majoring in Psychology, and I began playing in a three-piece rock band we called "Pressure." The members consisted of Chuck Zoubek (who later changed his last name to Beck) on lead guitar; my brother, Bruce, who had become quite an accomplished drummer, and me, on bass guitar. For a three-piece band, we sounded pretty good.

I graduated from Wright State University in 1974 with a Bachelor of Arts degree in Psychology and soon realized there was nothing I could do with that degree. I was offered a job as Foreman at International Harvester here in Springfield and the starting pay was $12,500 a year which wasn't too bad back then. Right after graduation, my good friend, Chuck McCall, and I took off out west in his van for about three weeks. We went to Colorado Springs to visit Judy Albin and Tammi; went to Alamogordo, New Mexico to visit Roger and Peggy Clem; went to Carlsbad Caverns and watched all the bats fly out at night, and then we traveled on down into Mexico. We camped on the rim of the Grand Canyon, drove through Monument Valley and spent the night at some pull-over in Utah. I had never in my life seen so many stars from horizon to horizon. It was a miraculous sight.

I began dating my future wife, Sharon, in 1974. We met on a blind date, and I later discovered that her parents also met on a blind date, as did my parents. Sharon was seven years younger than I was which means she was in Junior High School when I was in Vietnam. We've had to make some adjustments over the years, like any marriage, but we've lasted 49 years thus far.

There were many teachers in my family, so I decided to go that route. The GI bill helped me pay for my undergraduate degree and then there was a Vietnam Bonus Bill that helped me obtain my master's in education degree.

Sharon and I were married on March 22, 1975. I was working at Shillito Rike's at the new Upper Valley Mall in housekeeping, all the while attending WSU for my master's degree. Sharon was working as a dental assistant then for a Dr. Byers.

I graduated from WSU with a master's degree in education, in 1977. I began teaching at Town and Country School in January 1976, the year before a big blizzard. I was still playing in night clubs but burning the candle at both ends wasn't working for me. I had to quit playing in night clubs.

Nine years later, I was hired to teach at South Vienna Elementary and Junior High School, eighth grade special education. Nowadays special ed. teachers are called Intervention Specialists.

There were never any celebrations or ceremonies or acknowledgements for Vietnam vets, until one year at South Vienna School where I was a teacher, sometime in the early 1990's, a group of elementary students came into my classroom and one by one they came to me at my desk, and laid a card they had made on my desk, and told me, "Thank-you for fighting for our country." That was the VERY FIRST time that ever happened in my life. It was a Veteran's Day I shall never forget. I was naturally in tears, and I thanked them profusely. Since about that time, there has been an awareness raised about Vietnam and other veterans, in our schools and other public places. It is a good feeling now, and not a sad one.

I first met Roger Tackett, a Marine veteran and a Spring-field County Commissioner for over 20+ years, at South Vienna School. Our classrooms at the time were open space and he had been asked to speak to some students in a class next to mine. When I heard him mention Vietnam, I went over and listened to him tell the class about how he was wounded in Vietnam in 1967 and how a medic had treated him and saved his life. Roger was in a wheelchair, paralyzed from the waist down.

Roger will spend the remainder of his life in a wheelchair due to the wounds he received in 1967 from a sniper's bullet. As Roger was speaking, he looked over at me listening and he stopped. He stared at me for a second and said, "You were there, weren't you." Tears formed in his eyes. I told him yes, I was, but to go ahead and finish speaking to the students. We talked afterwards and have since become very good friends and remain so to this day. There is an unbreakable bond between veterans, especially combat veterans, and it is a bond that I respect and cherish.

At South Vienna School, I became friends with a Marine vet, Denny Holloway, and another Airborne Army Ranger vet, Rick Farmer. Both friends were wounded in action and are decorated heroes. Not many people know that. Rick was stabbed on a night patrol, a stab that was meant for his heart but went into his shoulder. Denny was stabbed multiple times after diving into a hole while his Marines were trying take a hill. In that hole was a North Vietnamese Regular who began stabbing Denny. Luckily someone came to Denny's rescue. When Denny came home to finish out his time in the Marines, Denny was assigned to a security detail at Arlington Cemetery.

Marine Denny Holloway

Marine Rick Farmer

On one occasion, Denny was part of a ceremonial memorial team charged with escorting a coffin to be buried at Arlington. There were protesters there awaiting the procession and Denny's captain ordered the honor guard not to react or move, no matter what. One of the protestors walked up to Denny, stuck flowers down the barrel of his weapon and commenced peeing on his leg. Denny did not move. I don't know what I would have done.

As I said before, my father told me that my memories of Vietnam would never go away, but would lessen in intensity over time. Mom told me later that Dad had confided in me things that he had never even told her. Dad and I bonded in a unique way, a way that had not existed before. I appreciated his counseling and his guidance, and for the time being, I felt better.

Commissioner Roger Tackett's wife, Martha, was the first person to tell me about how Agent Orange was affecting veterans who served in Vietnam and she was the first person to tell me about the veteran's clinic on Burnett Rd. This information was to change my life.

As time passed, I sought professional counseling at the local VA Clinic on Burnett Rd. My counselor was a very pleasant sociologist, but I was asked to recall many things I had tucked away very neatly for a long time, and often felt worse upon leaving a session than entering. Many times after leaving my session, I would drive to Rose Hill Cemetery and park and just sit in my car for a while.

I stopped going. My thinking was, and still is, just to let sleeping dogs lie. If I don't dredge these memories up, then I won't have to deal with them. I know I will hear shouts of protest from professional folks on my line of thinking, but I must do what I think is best for me. It may not be good advice for someone else.

I personally think that the trouble with "flashbacks" is that you relive a "then" event with a "now" brain, and what you were able to handle back then, you may have more trouble handling in the present. That is, I believe, what is so devastating to so many veterans.

There have been moments where I was caught completely off guard. When the Desert Storm troops began to return home, and the TV would show all the flag waving, bands playing, and cheering after a few months of combat, I began crying for what I, and thousands of other soldiers like me, had not experienced.

The WWII veterans involved in the Pacific War, and the Korean War veterans had no fanfare upon returning, but at least they were not greeted with jeers and spit and signs. I can't explain my feelings or my reaction. Maybe I was just feeling sorry for myself. Regardless, I was glad for the returning soldiers, but at the same time, sad that we were treated the way we were.

I was assigned a physician assistant at the Burnett Rd. Clinic, Earl Morse. Earl was a great and a very caring person, and you may recognize his name, for he was a co-founder of the Honor Flight Network, an organization that transports veterans to see their memorials in Washington D.C. at no expense for the veterans. He was moved by learning that very few WWII veterans had not or could not go to see the newly built WWII Memorial in Washington, D.C. In 2005, he enlisted some of his fellow pilot buddies and arranged for six private planes and pilots to carry two WWII veterans each to see the WWII Memorial. This was the beginning of a program that has grown enormously. Earl asked me if I would like to serve on the National Board of Directors for the Network which met every month to handle the particulars of the organization. I gladly accepted and served in that capacity for a few years.

I want to tell you that it was Earl Morse, when he found out I was a combat medic in Vietnam, asked me if I'd be willing to talk to a clinical psychologist about my tour in Vietnam as a medic. I asked him, jokingly, "Why? Do you think there's something wrong with me?" "No," he said, "I just would like to be sure that what you experienced over there wont be something that may disrupt your life later on." Earl made an appointment for me to see a psychologist at the Dayton VA, and I went. She was very nice and very thorough. She recorded everything, every knee bounce, and finger tap as we were talking. I thought the session went well but I was surprised that in two weeks, I was being compensated for PTSD.

My affiliation with Honor Flight and veterans, in general, have led me down a path I really appreciate and hold dear. I have made some good friends with fellow Vietnam veterans. I have made friends with, and visited regularly, WWII veterans. I still have recurring nightmares and flashbacks, but not to the extent that I once had. It has helped me emotionally just to write this, and for what it is worth, if anything good can come out of such an experience it is this: I appreciate every day of my life; I go to bed thankful for a roof over my head, and a pillow to lay my head on; I don't become upset over things that often cause other people stress; I realize how fleeting and uncertain life can be, which makes friendships and relationships something to be cherished and protected; and lastly, I feel that I want to spend the remainder of my life helping people, especially veterans.

For whatever reason, I was allowed to live through Vietnam, and I do not take that lightly. Many veterans have felt the guilt of survival, remembering those who fell in battle or to some other effect of war, such as Agent Orange. I went through those feelings, but after realizing there was nothing I could do to justify it or explain it, I decided to be grateful and useful in whatever I decided to do, never taking my life for granted.

I have been blessed many times over with a loving family, and a family of close friends. At the funeral of my father's cousin, Harry Ark, a man whom many people admired and respected, I remember one of Harry's favorite quotes: *"Never withhold your hand to help someone in need, for you do not live on this earth alone. Your brothers and sisters live here too."* Harry was a WWII veteran who served in the Pacific theater.

This quote helps me understand my purpose in Vietnam. I was there to help my comrades make it through. It also gives me a reason for my existence. I am here to serve mankind, and in doing so, I am serving God.

—Randy Ark

Because of my Vietnam experience, my circle of friends and acquaintances and associations produced an array colorful people. City leaders, County leaders, commanders of various military organizations, WWII veterans, Korean War veterans and a multitude of Vietnam veterans.

Below are two WWII veterans whom I became close friends with. Their experiences were extraordinary.

John Kunkel, 4th Infantry Division, 22nd Regiment

John Kunkel in 1944 (left) and 2007 (right).

While in high school, my daughter, Kara, was employed at a little restaurant on Derr Road, called Suzi's. One night she came home and told me that I just *had* to meet this older gentleman who came in frequently to eat in the afternoons. Knowing my interest in WWII veterans and certain battles in the war, she revealed to me that he was in the Battle of the Bulge. That was all it took. His name was John Kunkel, and I agreed to meet with him. This was the beginning of a long and endearing friendship.

I went many times afterwards, to visit John at his home in Northridge, and then to Oakwood Village (a retirement community), when he moved there. One morning, I picked John up in my van and we drove to meet other veterans at the Springfield Airport. When we arrived, we watched as a B-17G, a C-47, and a P-47 flew in and landed right in front of us. He really enjoyed that.

John was not reluctant to tell me about his time in the service, for he thought people needed to know what war was like, and to get information from someone who had been there, and John had definitely been there!

John entered Europe just after D-Day through Le Harve, France. He was then transported to St. Mere Eglise, and from there he went through St. Lo, the hedgerows of Normandy, and entered into the Hurtgen Forest on November 16, 1944. John told me that when he was at St. Lo, the whole place had been leveled by artillery shelling, and that nothing was left standing except an old church. I happened to locate that same church when I was in France in 2004, and there was an artillery shell still embedded in the side of the structure. It had never been removed. I took a picture for John.

The church in St. Lo, France

the artillery shell in the wall of the church in St. Lo, France

I knew a little about what happened on the beaches of Normandy, and the Battle of the Bulge, but up until this point, I had never even *heard* of the Hurtgen Forest! John was very emphatic when he told me that of all his experiences in the war, fighting in the Hurtgen Forest was, by far, the worst. I began to listen to his descriptions of the forest and what happened there. I also read articles on this battle that seemed to have eluded me in my studies of WWII. What I discovered simply astounded me.

John was in the 22nd Regiment of the 4th Infantry Division.

The Battle of the Hurtgen Forest, which began on Sept. 19, 1944, lasted a total 90 days and involved nine American Divisions and their supporting units. *More than 24,000 Americans lost their lives and there were another 9,000 casualties from trench foot, disease, and combat exhaustion.* John was wounded from an artillery blast on December 4th, 1944, but made it back to his unit in time to fight in *'The Battle of the Bulge'* in the Ardennes Forest.

John told me of a time (Dec. 22nd, 1944) when he and his patrol were watching a river crossing, using a French Chateau for shelter. John heard an artillery shell going over the chateau and then heard another in front of the chateau. He knew then that the next shell would be right dead-center on the chateau. Just as he was about to yell at his men to get out of the chateau, the shell hit. He was blown out into a snow drift, landing along the banks of the Moselle River, and all five of his men inside the chateau were killed.

There happened to be two medics just down the road and they saw the shelling and ran to find John in the snow. At first, they thought John was dead, but because one of his eyes was lying outside its socket and blood was pulsing from inside the socket, one of the medics realized he was still alive. John told me the medic simply licked his fingers, pushed his eyeball back into its socket, and secured it with a patch. His eye is working fine to this day.

Through his travels and experiences, John met several interesting people and one of those that he met was Ernest Hemingway, but he did not care for him at all.

John also told me that he was not only trained as a sniper, but he was a self-taught barber who would cut any soldier's hair that would pay him, so he became affectionately known as "the snippin' sniper".

John is a devout Christian man, who has authored a book, "The Confederate Yankee," and has written published poetry, and at one time built many intricate pieces of miniature furniture. He still has many pieces in his China cabinet that one may view.

John has earned two Purple Heart medals, two Bronze Star medals, a Combat Infantry Badge, and various other medals and citations.

It took John 47 years to reclaim his army records, because his records, along with others of his unit, were buried in the Hurtgen Forest, and were not recovered for years. They were buried because it was believed that they would soon be overrun by the Germans, and they did not want any personal records to fall into enemy hands.

John said there is one thing that he would like to get out of his mind; the horrible things he saw at Malmedy, Belgium, where what is known as the Malmedy Massacre occurred. He said that he hoped that history would never forget it, nor the 101st at Bastogne.

Malmedy was the site of a brutal German massacre that left 71 soldiers dead and 21 survivors. The soldiers were prisoners and unarmed and were murdered execution style.

John passed away in 2008 as one of my dearest friends, and I never tired of hearing his stories, or laughing at his wit and humor.

To many folks, John was just an old man in a retirement facility, Oak Wood Village, whom many people pass by, never knowing what an incredible man he is or what amazing experiences he has had.

For me, John will always be much, much, more.

Harold Deane, 3rd Infantry Division, WWII

John often spoke to me about a good friend of his, also a WWII veteran, Harold Deane, who had served in the 3rd Infantry Division, and wanted me to meet him. When John was hospitalized one time, I went to see him and noticed a man in the parking lot walking to his car that I thought might be Harold. I yelled out his name. It was!

Harold Deane
and John Kunkel

After John passed away, Harold, his wife Betty, and I, became very close. Harold and I went to Glen Haven Cemetery every Memorial Day weekend to visit John's grave and Harold would stand by John's grave and salute. Turns out, John's wife Betty knew my father when she was younger. Small world. Riding around Springfield with Harold was a history lesson for me. We would pass by old churches and other old buildings and Harold would say, "I remember when that church was being built."

Harold Deane saluting his buddy, John Kunkel, today at Glen Haven Cemetery. 5/28/11

At John Kunkel's viewing, after everyone had left, County Commissioner Marine Roger Tackett, Harold Deane, and myself pulled two folding chairs up next to John's casket and just sat and talked for a while. John was lying there in the casket, and we were sitting as close as if John were in a chair with us. It was quiet and the lighting was subdued. This was a very special moment for me.

Harold told me that when he was wounded, he was being treated near to where Audie Murphy was being treated under the same tent. He also got to meet Ernie Pyle and liked him very much. John said that he was often offered an increase in rank, but he would always refuse. He did not want to be responsible for other men's lives.

Harold served overseas with the Third Infantry Division, away from his home here in Ohio for three years and three months. He participated in six beach-landings coming off Higgins boats, beginning with "Operation Torch" in 1942 on the northern coast of Africa. Two more landings in Sicily, two on the shores of Italy (Anzio, being one of them) and the last one in Southern France. You may already know this, but James Arness (Gunsmoke's Matt Dillon) was wounded on Anzio.

Harold said that he had a hard time sleeping when he came home. His wife, Betty, told me that when he first came home from the war, he slept on the hard wood floor next to their bed.

Harold's wife, Betty, also told me, "When Harold went into the Army, he didn't drink, smoke, or cuss. When he came back, he did all three." Made me laugh.

In 2010, Harold received France's highest Award, **The Knight of Legion of Merit Award** given to him by Governor Strickland at a ceremony in Columbus, Ohio. I took Betty and him to this event at the Capital Building.

Harold receiving the Knight of Legion of Merit Award

Harold was proud of his service and did not mind telling his stories. (over and over). Betty would patiently and quietly listen every time, unless Harold got stuck, then she would refresh his memory.

Harold had many interests, and many different occupations in his lifetime:

- He was a Springfield City bus driver
- He got his pilot's license and learned to fly
- He was a reserve police officer
- He was a special deputy sheriff
- He was a rural route mail carrier
- He served with the volunteer fire department
- He was a church member / Clark Lodge 101 Mason

Harold Deane
1935 Golden Glove
Welter Weight Champion
Springfield, Ohio

WWII

3ʳᵈ Infantry Division
10ᵗʰ Combat Engineers

June 1942 to Sept. 1945

1935 Golden Glove Welter Weight Champion of Springfield

He attended Third Division reunions, and Chapter 620 meetings.

Harold loved his family, but he had nearly outlived them all! Harold was 97 years old. Born in 1915. He stated once that he had no idea why God would want him around this long. I told him, "Harold, God is keeping you around for me!" Harold and Betty, both told me once that I was like a son to them and it's hard to describe what that meant to me. Harold surprised me one time by gifting me his military watch that was made for him. The face of the watch had the dates Harold was wounded in St.Di, France and his name.

Harold also repeated to me many times how he hated the way Vietnam veterans were treated when they came home.

When Harold died, I was asked to perform his eulogy which I gladly did.

It was Harold who secured me a place on the Purple Heart Float in the Springfield Memorial Day Parade by asking our then

Commander Doug Wood if it would be ok if I rode, after explaining my Purple Heart situation. I had confided in Harold early in our relationship about being wounded and not receiving a Purple Heart and the circumstances surrounding that. I didn't know anything about Springfield's Purple Heart Chapter 620 at the time and never had any hopes of getting my papers to qualify me for the medal.

I rode the float in the Memorial Day Parade and waved to the people along the parade route and afterward I was told that I could attend their chapter meetings, but since I had no papers to verify that I had earned a Purple Heart, that I wouldn't be able to vote on anything, because I was not an official member. But they did believe what I told them about being wounded.

At the time I began attending the monthly meetings, the Chapter 620 was in the midst of trying to raise money for a War Dog Memorial that would be placed into Clark County's Veterans Memorial Park, so I went to work to help them raise money for the memorial.

The reason for this particular memorial was that when Commander Dave Bauer was wounded in Vietnam, there was a wounded war dog that was brought in on a stretcher and placed in the surgical tent next to where Dave was. This occurrence made an impression on Dave and made him want to honor war dogs in some way. I always teased Dave about the physicians attending to the dog first, before they treated him.

I was saddened to find out that these war dogs were euthanized after they had served their purpose and could not be brought home because of the way they were trained. I witnessed a LLRP veteran (long range reconnaissance patrol) walk out from the jungle next to our battery one time with a war dog on a leash. The dog was wearing a muzzle which was required when around anyone but his handler. I am thinking that anyone who worked with and served with these war dogs would have a different story to tell about their loyalty, etc.

After our chapter had their war dog memorial constructed and placed at the Clark County Veterans Memorial Park, they had a dedication ceremony, and among those in attendance were City Police Canines and Highway Patrol Canines. Many local folks attended including Springfield's Mayor, Warren Copeland, and many law enforcement officers.

Trooper Tammy Getz with her canine partner, "Ali," and Officer Mike Fredendall with his canine partner, "Rambo." Thank-you from Chapter 620 M.O.P.H.

Dave Bauer and Doug Wood approached me one day about helping them get Springfield's Fountain Avenue sub-named "Purple Heart Way." Fountain Ave. went right next to Clark County's Veterans Memorial Park and was also used as the Memorial Day Parade route. Fountain Ave. is one of the main streets in Springfield. They were told at the County Office that they needed the names and addresses of every dwelling along Fountain Ave. so that those residents could be notified of their street possibly being sub-named Purple Heart Way.

I enlisted the help of my longtime friend, Charles Swaney whose law office was located along Fountain Ave. and he had his secretary type up the list we needed.

We then needed to approach the City Commissioners to get approval for what we wanted to do and ask them to finance ½ the cost. After speaking with everybody who was anybody and collecting signatures showing support for our project and after I addressed the Springfield City Commission on three separate occasions, they agreed we could go ahead with our project and would pay half the cost of the new street signs and two backlit signs that would be placed on Main Street and High Street. They submitted on the condition that our chapter would come up with the other half. I assured them that we would come up with our half. Fountain Ave. was officially to become sub-named, "Purple Heart Way."

On the morning when the new signs were to be installed, many of the County and City Commissioners, curious town folk, Mayor Copeland, and the Clark County Sheriff, Gene Kelly, were there. I was very proud to be a part of this endeavor.

Dave Bauer had another project in mind, and he now wanted to create what he called a Dog Tag Memorial, to acknowledge and honor every person from Clark County who had died in Vietnam and have their names etched into a granite memorial and put on display in Clark County Veterans Memorial Park. In the meantime, Dave was made State Purple Heart Commander which left me to try and raise the money for the new memorial and I also had to gain permission from the folks who oversee the park grounds so we could place our memorial there.

I attended many meetings with civic organizers, city and park officials, VFW's, Moose Lodges, American Legions, and the Veterans Commission, etc., to gain permission and to raise money. A former Greenon High School classmate surprised me when he generously donated a large sum of money that got us on our way.

Mike Spradlin graciously created the dog tag chain for our memorial. He donated his time and talent and saved us thousands of dollars in doing this. Mike is Vice-President of Spradlin Bros. Welding Co. and a good Christian man.

There were many people in attendance at our dedication, and many people expressed their gratitude and heartfelt appreciation for creating this lasting memorial to their loved ones. And some shed a few tears seeing their loved one's name etched in stone for all to see and be witness to for years to come. For every person who was there who was associated with any of the names on that memorial, we had made up a set of dog tags on a chain for each of them. I was so glad we did this! This was September 2015.

Dedication of Dog Tag Memorial

New memorial at Veterans Park

Three of the nearly 300 people who attended Saturday's dedication of the Vietnam Dog Tag Memorial at Veterans Memorial Park in Springfield examine the new monument. It honors and acknowledges the 63 men from Clark County who sacrificed their lives in the Vietnam War. Donations from individuals, The Turner Foundation and a couple of veterans groups helped fund the monument, which was two years in the making.

Clark County Veterans Memorial Park
August 12, 2020

History was made today!

Today, Chapter 620 M.O.P.H. purchased two QR Code
plaques to be mounted on the Purple Heart Monument
downtown and the War Dog Memorial located at
Veterans Memorial Park.
Pictured above from left to right:
Doug Wood, Jim Ryan, Randy Ark,
and Josh Walters of Dodd's Monuments.
02/21/12

* these purchases are the first in Clark County for
Civic or Personal Memorials

With the help of Josh Walters of Dodd's Memorials, our chapter purchased a QR code for each of our memorials. Each code contains information and pictures on the construction and placement of the memorial and pertinent information on the various veterans' organizations located in the Springfield, New Carlisle, and South Charleston areas, along with contact information of said areas.

Donations were coming in from unexpected sources after this dedication. There was an elderly woman who lived in Enon, Ohio, who gifted me enough money where I could start a foundation that would provide for the upkeep of the park in the future. Marine Ronnie Coss and I were asked to attend a meeting with her lawyer to work out the details. She remarked at the meeting, "I immediately knew I could trust you." She had been watching what I had been doing here in Springfield and trying to do in the Clark County Veterans Memorial Park and appreciated my efforts and decided to help in some way. She told me that her husband, who was deceased, had flown a B-26 bomber on the morning of D-Day in WWII. I try to keep in touch with her.

Another surprise donation came at the beginning of our project from a classmate who was a few years ahead of me at Greenon High School. His donation went a long way toward purchasing our Dog Tag Monument from Dodds Monuments.

Clark County Veterans Memorial Park
August 12, 2020

For two years, the welding program at Clark County Vocational School donated money earned by auctioning off items they created in welding class. The money was designated to help with the renovation of the Clark County Veterans Memorial Park.

Another classmate who lives in California donated a substantial amount of money, enough that allowed us to purchase five military flags and poles, installed, with mounted solar lights.

2019
Clark County Veterans Memorial Park

The Harley-Davidson Owners Group, Springfield Chapter HOG #3872 donated enough to purchase a 12" x 12" paver for the veterans' walkway through Veterans Memorial Park.

One of my favorite activities as a veteran was speaking to school children and Snowhill School here in Springfield had a History Club headed by teacher Larry Marple who invited me and Dave Bauer to come and speak to the club and good friend, Mike Spradlin, Vice-President of Spradlin Welding, would sometimes bring Army vehicles that he had refurbished so the kids could experience those vehicles first hand. Here is an article I wrote for our local newspaper after one of our visits.

Vietnam Vets Bring Information To All, Comfort To One

By RANDALL W. ARK
Contributing Writer

Purple Heart Commander Dave Bauer and I were invited, once again, to speak to the History Club at Snowhill Elementary School located on the corner of Harding Rd. and St. Paris Rd. Teacher Larry Marple's History Club consists of students ranging from the 4th through the 6th grades. In addition, Mike Spradlin and his associates, Jim and Jeff Hazlett, displayed Vietnam era military vehicles explaining their use and importance during the Vietnam conflict.

After we had finished speaking, and the students were allowed to examine the vehicles, one student stood off from the others. An adult there brought him to my attention explaining that the student was having a difficult time thinking about his grandfather and asked if I would speak to him. A teary-eyed young man came over to the bench where I was seated and sat down next to me.

I learned from him that his grandfather, whom he was named after, had suffered from the effects of Agent Orange and had passed away before this young man was born. He was grieving over a grandfather he had never met.

Commander Bauer and I spoke with him at length on how all those who served in Vietnam were considered "broth-

ers" and that his grandfather was a brother to us, and a true patriot for serving his country. We told him that many veterans were affected by Agent Orange and that I had trouble walking, because of it. We also told him that one of the reasons we (the Commander and myself) speak to students and other groups is so that the men and women who served this country, like his grandfather, would not be forgotten. His grandfather should be remembered as a man who fought for others when duty called. We both assured him that we would not let his grandfather's memory die. In addition, I told him where he might learn more about

his grandfather's service and where to find that information.

This young man thanked us for speaking with him about his grandfather and that he felt a better knowing that he was not alone in remembering his grandfather's service and sacrifice, that it meant a lot to many people.

The war in Vietnam has affected the lives of people in many different ways over the years and continues to do so. On May 8th of 2012, the war continues to affect a young boy at Snowhill Elementary School. We owe a debt of gratitude to the men and women like this young boy's grandfather who sacrificed and served when duty called.

All the while, every member of Chapter 620 kept encouraging me to keep applying for my Purple Heart. When I received a letter of denial from the state board of examiners and was about to give up, I get a call from the Dayton VA. I was standing on the deck of a docked destroyer in Louisiana at the time. I recently had x-rays on my back at the VA Hospital and they asked me how I might have gotten pieces of metal around my backbone. I told them it was probably shrapnel from a mortar or rocket explosion in Vietnam. I was cautioned not to have any MRI's for the next three years for fear the metal would be pulled out through the skin by the magnetism of the machine.

I applied to the state board one more time for a Purple Heart verification thinking that the papers from the hospital stating that I had shrapnel in my back would make a difference in their decision to grant me Purple Heart verification. It did not. I was told in a letter that since I had already applied once, they weren't even going to look at my new submission, the one that had the shrapnel information. Well, that didn't seem right.

It seemed like I was beating a dead horse, so I resigned myself to accepting my situation the way it was and just go forward.

I continued helping our chapter with the renovation of our Clark County Veterans Memorial Park. By working on renovating our park, I was able to connect with people whom I would not have otherwise. One of my contacts provided me and some fellow Chapter 620 members access to where the refurbishing of the Memphis Belle was in progress.

The Memphis Belle
November 8, 2012

Randy Ark, Dick Haines,
Doug Wood, Dave Bauer,
and Lew Waters

Chapter 620 M.O.P.H.

If you recall, I mentioned earlier that I was awarded an Army Commendation Medal with a "V" device for actions in Vietnam. I was notified that that medal qualified me to be inducted into the Ohio Military Hall of Fame for Valor. The induction ceremony was held in the Capital Building in Columbus, Ohio. This was in April of 2015. Family, friends, and fellow veterans attended as well as local TV stations.

Ohio Military Hall of Fame for Valor
Columbus, Ohio 4/24/15

Lapel Pin

After years of denials and rejections and submissions, I was approached by a Mr.Ben Thaeler at one of our memorial ceremonies, who worked for and with Congressman Warren Davidson. After speaking with Ben, he suggested that I give him copies of all my papers including my Dayton VA Hospital records where they reported shrapnel in my body and who had advised me not to get an MRI for three years.

I had given up hope of ever getting my Purple Heart medal,

but Congressman Davidson's office encouraged me all along the way for two years. One day, a UPS driver appeared at my front door with a package to sign for. I opened the package, thinking it was probably medicine from the VA and I discovered a Purple Heart Medal with my name on it! I was dumbfounded and couldn't believe it actually had happened.

I was wounded in February of 1969, and it had been 16 years since I first applied for verification and here the medal was in my hand. I couldn't wait to tell the members in Chapter 620. Now I could vote and hold office and wear my Purple Heart Blazer at special ceremonies. Now, I was bonified!

Springfield veteran's patience pays off with Purple Heart medal

I had a wonderful opportunity to thank Congressman Warren Davidson and Mr. Ben Thealer in person at our new Community Based Outpatient Clinic (CBOC). I will never forget the work they did on my behalf.

Things are winding down in our chapter here in 2024 Springfield, as well as other veterans' organizations. Older members have died off and there seems to be little interest in the more recent veteran population to join veterans' groups. We are all in our mid to late 70's, and Agent Orange has taken its toll on many Vietnam veterans and I've read that we are dying at a faster rate than WWII veterans. As far as gaining members and having fund raisers, we are sadly at a standstill.

Regardless, I am glad to have lived as long as I have. I've been blessed in so many ways with family, close friends, and veterans' groups. The Lord has blessed me with many opportunities to assist veterans in getting the healthcare they deserve, and the compensation that may be due to them.

My experiences and my time serving our country has shaped my character, my morals, and my patriotism. It has been a rewarding journey as well as a painful one at times, but in retrospect I am glad to have had the experiences of both.

I hope there were things in this book that one could relate to and hopefully benefit from. It is also my hope that Americans will continue to honor and take care of the men and women who have served our country and are serving today. Our freedoms were bought with a price and for as long as I live, I will work to ensure that people don't forget all those who sacrificed in the service of our country and our way of life.

Additional Images

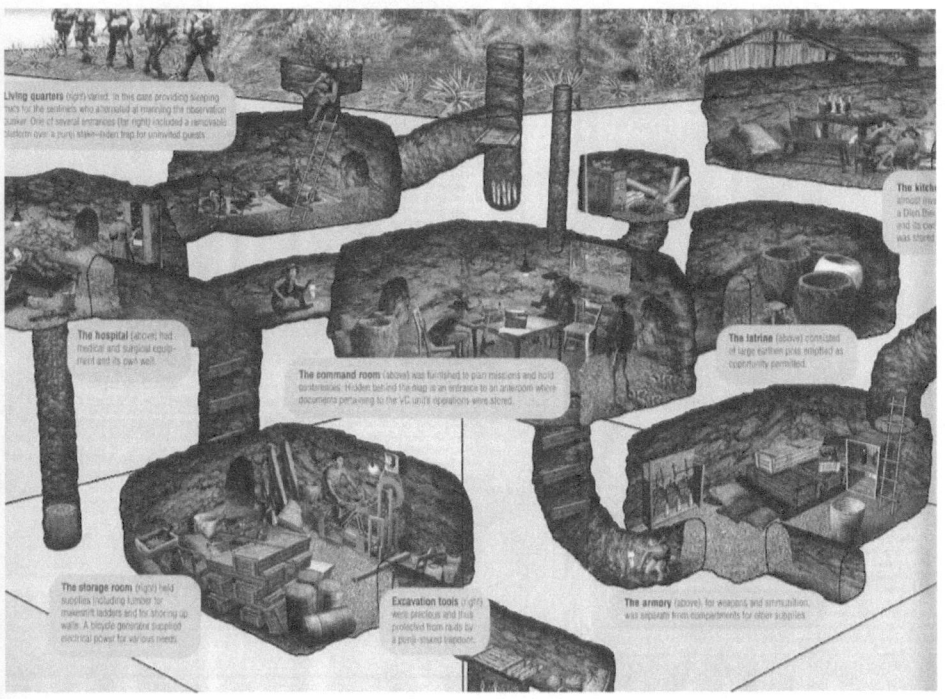

1st Battalion,
3rd Marine Regiment
Memorial

Dedicated
July 3, 2016

The person pictured with me above is Joseph Galloway, the author of, "We Were Soldiers Once, and Young."

My grandaughter, Morgan, my wife, Sharon, and I, are looking over what will be, the "Dog Tag" monument. May 28, 2014

Some of the street banners I made on computer and
were hung on Fountain Ave. and Cliff Park Rd.

1915 Harold Deane 2012

Dedication of Dog Tag Memorial

Lai Khe, Vietnam 1969

Private First Class Randall W. Ark

Class of 2015

United States Army

Army Commendation Medal with "V" Device

For heroism in connection with military operations against a hostile force in the Republic of Vietnam: On this date, Private First Class Ark was serving as a medical aidman with his battery when at approximately 0245 hours, the friendly base was suddenly subjected to an intense barrage of enemy mortar and rocket fire. During the initial volley of hostile fire, a casualty was sustained in one of the howitzer sections. Without regard for his personal welfare, Private First Class Ark braved the vicious fusillade of shrapnel and the impacting enemy rockets as he maneuvered from the relative safety of his bunker to the wounded man's location, and administered first aid. When eight men were seriously wounded in another howitzer section, Private First Class Ark again exposed himself to the aggressors' barrage as he rushed to aid his fallen comrades. His selfless courage and professional performance of his mission served as an inspiration to the men of his unit and earned him the respect and confidence of all who witnessed his actions. Private First Class Ark's actions are in keeping with the finest traditions of the military service and reflect great credit upon himself, the 1st Infantry Division, and the United States Army.

Acknowledgements

I have thought a long time about writing this memoir about my time in Vietnam. I thank my Family and my friends who encouraged me, and the Deeds Publishing team for making it a published reality. I encourage other veterans to stop putting it off and write your memoir.

About the Author

Randy Ark was raised on a small farm outside Springfield, Ohio. He was drafted into the Army in 1967, a year after high school graduation. He served as a medic in Vietnam with the 1st Infantry Division from August 1968 to August 1969, where he was awarded three Army Commendation Medals and a Purple Heart.

He earned a Master's Degree in Education from Wright State University and taught school for 32 years, retiring in 2007. After retiring, he became involved with the Military Order of Purple Heart and other veterans organizations. He serves on the Board of Directors of the National Honor Flight Network.

Randy is married with three grown children and four grandchildren.

www.ingramcontent.com/pod-product-compliance
Lightning Source LLC
Chambersburg PA
CBHW020357130626
46549CB00006B/2311